Steam Boiler Operation:

principles and practice

SECOND EDITION

JAMES J. JACKSON
Plant Engineering Manager
Chemical Group, Engelhard Division
Engelhard Minerals & Chemicals Co.

Prentice-Hall, Inc., Englewood Cliffs, New Jersey 07632

Library of Congress Cataloging-in-Publication Data

Jackson, James J., 1926–
    Steam boiler operation.

    Includes index.
    1. Steam-boilers.   I. Title.
TJ286.J32   1987   621.1′94   86-30439
ISBN 0–13–846346–8

Editorial/production supervision: *Raeia Maes*
Cover design: 20/20 Services, Inc.
Manufacturing buyer: *Rhett Conklin*

Printed in the United States of America

10 9 8 7 6 5 4 3 2 1

ISBN 0-13-846346-8   025

Prentice-Hall International (UK) Limited, *London*
Prentice-Hall of Australia Pty. Limited, *Sydney*
Prentice-Hall Canada Inc., *Toronto*
Prentice-Hall Hispanoamericana, S.A., *Mexico*
Prentice-Hall of India Private Limited, *New Delhi*
Prentice-Hall of Japan, Inc., *Tokyo*
Prentice-Hall of Southeast Asia Pte. Ltd., *Singapore*
Editora Prentice-Hall do Brasil, Ltda., *Rio de Janeiro*

Contents

PREFACE TO THE FIRST EDITION *vii*
PREFACE TO THE SECOND EDITION *ix*

1
A BRIEF HISTORY OF STEAM *1*

2
TEMPERATURE AND THE EFFECTS OF HEAT *17*

Temperature Measurement, *17*
Heat, *20*
  British thermal unit, 22
  Specific heat, 22
  Latent heat, 22
  Conduction, 23
  Convection, 23
  Radiation, 24
Effects of Heat, *24*
Effect of Heat on Gases, *25*
  Condensation, 29

3
FUELS *37*

Coal, *38*
Oil, *41*
Gas, *52*

4

BOILERS   *67*

Evolution, *67*
Stress in Shells and Joints, *79*
Types of Boilers, *89*
Care of Boilers (waterside), *93*
  Oxygen corrosion and pitting, 93
  Scale formation, 94
  "Wet" steam or carryover, 95
  Wetback and firebox, 96
  Gaskets, 97
Care of Boilers (fireside), *97*
  Furnace, tubes, and tube sheets, 97
  Tube cleaning, 98
  Gaskets, 98
Boiler Efficiency Measurement and Optimization, *99*
  Improving efficiency, 100

5

BOILER ACCESSORIES   *105*

Safety Valves, *105*
The Water Column, *111*
Globes, Gates, and Checks, *115*
Soot Blowers, *125*
Boiler Blowdown, *126*
Burners, *129*
Coal Burning, *132*
Feedwater Systems, *133*

6

BOILER AUXILIARIES   *139*

Pumps and Pumping, *139*
Pump Packing, *149*
Sealing Pumps, *153*
Pumping Problems, *160*
  Problem and possible cause, 161
Oil Fuel Systems, *163*
The Reducing Valve, *169*
Feedwater Heater, *171*

7

STEAM TRAPPING *179*

8

WATER TREATMENT *191*

External Treatment, *202*
Filtration, *202*
  Types of filters, 203
  Backwashing, 204
  Operation, 206
  Deaeration, 207
Softening and Demineralization, *209*
Demineralization, *211*
Mixed-bed Demineralization, *212*

9

INSTRUMENTS AND CONTROLS *219*

Temperature Measurement, *219*
  Bimetallic element, 220
  Filled systems, 220
  Thermocouple, 220
  Resistance pyrometer, 222
Pressure Measurement, *223*
Flow Measurement, *225*
  Steam flow, 225
  Btu meters, 226
  Liquid flow, 227
Boiler Water Level Control, *228*
  One-element control, 229
  Two-element control, 229
  Three-element control, 229
Combustion Control, *230*
Stack Gas Measurement, *232*

10

MANAGING THE POWER PLANT OPERATION *237*

Boiler Inspection, *239*
Closing the Boiler, *242*
  Raising steam, 242

Cutting In a Boiler, *243*
Laying Up the Boiler, *243*
  Wet layup, 244
  Dry layup, 244
Setting Up a Shift Schedule, *244*

11
BOILER ROOM SAFETY   *249*

Boilers, *249*
Pumps and Heaters, *251*
Fire Safety, *251*

GLOSSARY   *255*

INDEX   *267*

Preface
to the First Edition

Steam Boiler Operation: Principles and Practice was written to fill a need for a practical text on effective power plant operation. Present literature is either very elementary or assumes a college education. This book will provide practical information for the professional or the lay person.

Based on over 35 years experience in the production and use of steam, this book attempts to pass along the essence of that experience.

Before the advent of the "energy crisis," the operation of steam-producing equipment and the ultimate use of steam, for any purpose, was of little concern to the general public. Whether they be apartment dwellers, or home owners who rely on a boiler man for heat and hot water, or the head of a large corporation who assumes that somebody is looking after the power house, consumers of steam power have only recently begun to appreciate the role of steam energy in their daily lives. It has now become apparent that energy-conversion equipment such as boilers and engines must be operated efficiently. This book, therefore, is prepared as a guide for persons whose daily work is energy conversion—whether they operate a small heating boiler or the largest power plants.

Conservative estimates suggest that if all steam-producing and using equipment in the United States were operated efficiently, an immediate reduction of energy consumption of 25% would be possible. Rising costs of fuel and the need for this country to become as energy-independent as possible demand that these savings be achieved. It is to specialists in the power plant field that this nation must look for energy conservation.

JAMES J. JACKSON

Preface
to the Second Edition

The energy crisis of the past decade has made industrial, commercial, and domestic consumers of all forms of energy aware of the finite nature and high cost of nonrenewable fossil fuels. The failure of the nuclear power industry to establish itself as a viable alternative to fossil-fired plants assures, for the forseeable future, a reliance on coal, natural gas, and oil to fuel our power plants.

The basic laws of steam boiler operation remain valid, but in preparing the second edition, I have included the most widely used innovations that have been generated by the energy crisis. Due to the broad acceptance of the first edition, I have also had the benefit of suggestions from the academic world and power plant personnel, whose contributions have been incorporated in the revised text.

As in the past, it is the knowledge and skill of the watch engineer that will enable us to operate boilers efficiently to extract maximum value from all energy sources.

JAMES J. JACKSON

A Brief History of Steam

Lord, send a man like Robbie Burns, to sing the song of steam.

Rudyard Kipling, "M' Andrews Hymn"

The dictionary defines steam as "the gas or vapor into which water is changed by boiling, especially when used under pressure as a source of energy." Man has been familiar with steam for thousands of years, but only in the last two centuries has he discovered how to utilize steam to his advantage.

The application of early technology in the ancient world was influenced by the socio-economic structure of the era. As long as thousands of slaves could be used to do heavy work and supply the needs of a minority, there was no incentive to seek mechanical substitutes for human toil. The great Greek and Roman engineers had a remarkable knowledge of the properties of steam and hot air but made no attempt to apply their knowledge.

Hero of Alexandria was able to open the temple gates at sunrise by an ingenious arrangement of lenses concentrating the sun's rays on a tank of water that expanded when heated and caused the gates to open. His "Whirling Aeolipile" (see Fig. 1-1) employed in its simplest form the principle of the reaction turbine and jet engine of today, but at the time was considered only an amusing toy.

Figure 1-1. Hero's aeolipile, c. A.D. 50. *(Courtesy of British Crown Copyright, The Science Museum, London)*

After the Western Roman Empire collapsed in the fifth century A.D., Europe sank into the Dark Ages and no significant scientific inventions were produced for a thousand years.

The dawn of the Renaissance in the 15th century witnessed great progress in art, literature, and a wide range of learning, and set the stage for men to regard scientific inquiry and invention as essential for human progress. Many experiments were undertaken, which demonstrated a multitude of scientific principles.

In 1606, Giovanni Battista della Porta described two laboratory experiments that showed the power of steam to shift water by forcing the steam into a sealed tank of water. The steam built up pressure and forced out the water.

In della Porta's second experiment a flask with a long slender neck was filled with steam and the neck was inserted into a tank of water. The steam condensed, creating a vacuum and atmospheric pressure forced the water into the flask.

Soon after della Porta's experiments, Salomon de Caus, a Frenchman living in England, devised an ingenious use for steam pressure. de Caus, a noted landscape gardener, placed a spherical water tank over a fire, thus making a simple boiler. A pipe atop the sphere emitted a stream of water when the steam pressure was intensified by heat from the fire, thus creating a steam-powered fountain.

One of the great scientists of this era was Galileo, famous for his telescope, the pendulum, and his experiments with gravity. In 1641, he was consulted by the engineers of the Grand Duke of Tuscany after they had tried unsuccessfully to make a suction pump. The dukes' engineers were attempting to draw water from a depth of 50 feet. Galileo realized that pumps would draw water up from 28 feet and no deeper. He started experimenting to solve the problem, but died the following year.

One of his pupils, Evangelista Torricelli, continued Galileo's experiments and, in 1643, discovered that the pressure of the atmosphere would support a column of water 32 feet high if the upper end of the tube was sealed and all air pumped out. Torricelli and an associate, Vivani, went a step further and showed that atmospheric pressure would also support a column of mercury thirty inches high, thus demonstrating the principle of the simple barometer.

About this time, in the 1650's in Germany, Otto von Guericke was also experimenting with a vacuum and atmospheric pressure. He demonstrated the power of combining a vacuum with atmospheric pressure by filling a copper sphere with water and pumping it out. He created such a good vacuum that the atmospheric pressure crushed the sphere.

A later experiment by von Guericke was a valuable contribution towards development of the steam engine. He made a cylinder with a tight-fitting sliding piston. Twenty men hauled the piston to the top of the cylinder and the air was evacuated from the cylinder by connecting it to a vacuum sphere. The 20 men strained unsuccessfully to hold the piston at the top of the cylinder; but because of the atmospheric pressure above and the vacuum below, the piston slid down very easily.

It was now becoming clear that a powerful mechanical device could be built, if only a vacuum could be produced rapidly. della Porta had shown that water could be sucked up into a flask by the condensation of steam. Several inventors developed devices based on this principle.

Further work was carried out by Robert Boyle, an Irish chemist and Robert Hooke, his English assistant. Both men made significant contributions to the theory of heat engines, now known as thermodynamics.

Many experiments were being carried out at this time using steam pressure or atmospheric pressure combined with a vacuum. At the end of the 17th century a steam pump was built that effectively combined both sources of power. This was the work of Thomas Savery (Fig. 1-2) who, in 1698, patented an "engine for raising water by the impellant force of fire." A pump was positioned about halfway up a pipe between inlet and outlet. Steam from a boiler was piped to a closed tank until the tank was filled with the steam; then the supply was cut off and cold water was poured over the tank. The steam condensed, creating a vacuum which drew water up the suction pipe into the tank. A nonreturn valve in the suction pipe prevented the water from escaping back down the pipe. The steam cock was opened again and steam entered the tank, the pressure built up and forced the water through a second nonreturn valve to an outlet pipe. The cycle was then repeated.

There is no doubt that Savery's pumps worked, but they were little more than experimental prototypes. In 1699 he installed twin pumps working alternately on the banks of the Thames River in London. The necessary hand operation of the valves was a handicap, but the chief reason for his limited success was the inability of the pumps to lift water more than about 50 feet. The first stage, with a perfect vacuum, should have lifted the water 32 feet, but 20 to 25 feet was their limit and the pressure in the second stage was restricted by the boilers. Savery's boilers were dangerous, since they had no safety valves.

Savery's greatest contribution to the utilization of steam was probably his practical work improving valves and boilers. His predecessors had been experimental scientists and dreamers.

In 1663 the engineer Thomas Newcomen was born in Devonshire, England. He spent his early years in Devon and was apprenticed to a blacksmith and ironmonger. About 1685 he went into partnership with John Calley to supply Devonshire and Cornwall tin mines with needed tools and hardware.

Newcomen was acutely aware of the need to keep the mines free of water. Until this time, the mining of the tin had been basically

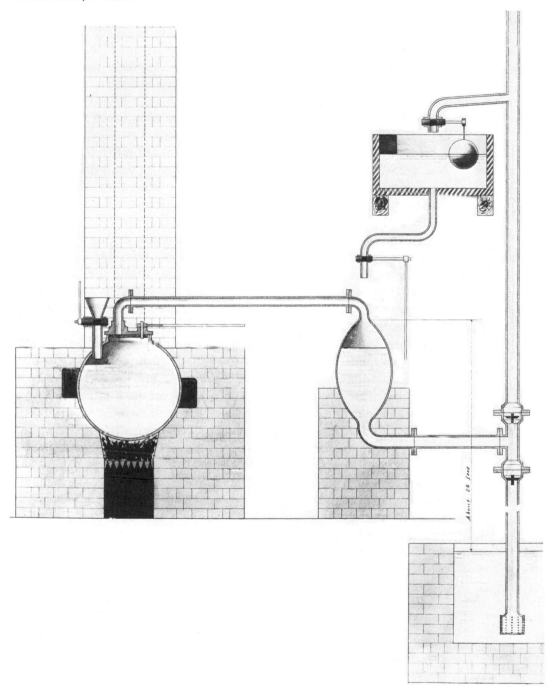

Figure 1-2. Savery's single pumping engine, c. 1699. *(Courtesy of British Crown Copyright, The Science Museum, London)*

surface mining, but with the surface deposits gone, shafts were being sunk to mine at subsurface levels. Being a metal worker, and familiar with all metals, he conducted experiments with power machinery at about the same time as Savery. In 1705, Newcomen and Calley entered into a partnership with Savery.

In 1712, after years of experimentation, Newcomen and Calley built their first successful steam-powered engine (see Fig. 1-3). It was a huge contraption some 30 feet high. Dominating the arrangement was a 25-foot-long oak beam, pivoted at the center, which rocked like a giant see-saw. The beam was used to transmit power from the engine's single cylinder to the water pump. It was, quite naturally, called a "beam engine."

The hand-operated mechanics of Newcomen's engine were based on Savery's original ideas. Under one end of the beam was a brass cylinder and beneath this a boiler. The water inside the boiler was heated from below by a coal fire. Inside the cylinder was a piston that moved up and down, a movement transmitted to the end of the beam by a chain. The other end of the beam was attached to rods that were, in turn, attached to a water pump. The pump end of the beam was made heavier so that the pump stopped in the "down" position and the cylinder end was in the "up" position with the piston at the top of its stroke. A valve was opened on the boiler, admitting steam to the cylinder. The cylinder filled with steam and the valve was then closed. A water-injection cock was opened, condensing the steam in the cylinder and creating a vacuum. Atmospheric pressure forced the piston down and the movement of the beam raised the pump, forcing water out of the mine. When the piston reached the bottom of the cylinder, the heavier end of the beam raised the piston, ready for the next working stroke. Steam was allowed into the cylinder on the upstroke and helped to drive out the water remaining from the condensed steam.

The valves and cocks on Newcomen's early engines were all operated by hand, usually by a boy who constantly had to tend the engine. Legend has it that an inventive young man named Humphrey Potter devised an automatic system of strings and cords to open and close the cocks, basing the device on the rocking movement of the beam. In fact the system of rods and levers later devised to open the cocks automatically was called the "Potter cord."

A "beam engine" therefore, is the basic mechanism that uses direct steam pressure or atmospheric pressure in conjunction with condensing steam. Condensation taking place within a cylinder using

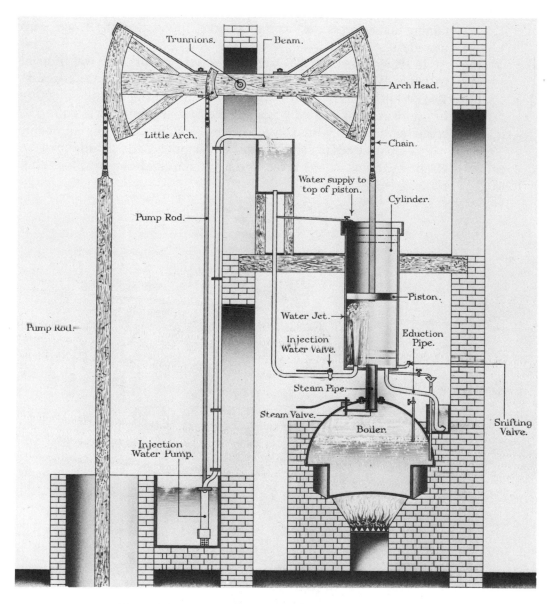

Figure 1-3. Newcomen's pumping engine, c. 1712. *(Courtesy of British Crown Copyright, The Science Museum, London)*

a water spray device was the main feature of what is called the Newcomen type engine. Some improvements were made by other engineers but the basic design remained the same for a century.

By 1725, the atmospheric pressure engine was firmly established and further progress awaited a more reliable boiler capable of sustaining higher pressures, as well as other innovations by a man with new ideas.

In 1764, a model of a Newcomen-type engine (Fig. 1-4) was brought to James Watt an instrument maker at Glasgow University (see Fig. 1-5). While repairing the model, Watt realized that a large quantity of heat was being wasted whenever the heated cylinder was cooled to create a vacuum. Watt solved the problem by designing an engine with a separate condensing device. He theorized that steam, being an elastic medium, would surge into a vacuum and if a pipe was con-

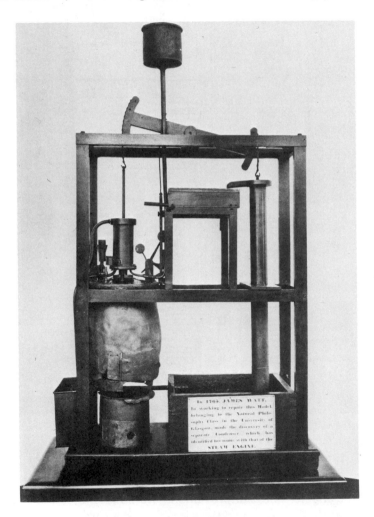

Figure 1-4. Model of Newcomen engine repaired by James Watt. *(Courtesy of The Science Museum, London)*

Figure 1-5. James Watt, 1815. *(Courtesy of The Science Museum, London)*

nected to a separate vacuum from the cylinder, the steam would rush in and be condensed without cooling the cylinder.

Like all steam engines, Watt's design had a cylinder and a piston but, in addition, a separate condenser and an air pump. The success of this operation depended on the air pump which sucked air out of the condenser, creating a partial vacuum. Steam from the cylinder rushed into this condenser, which was surrounded by cold water, and immediately condensed. The air at atmospheric pressure then moved the piston. Watt obtained a patent on his engine in 1769 (see Fig. 1-6).

The new method was first utilized when Watt became a partner of Matthew Boulton, a factory owner in Birmingham who was a shrewd businessman (Fig. 1-7). Watt refined his condensing engine, and by 1775 orders were received for the production of two of them.

The separate condenser (Fig. 1-8) was only one of many improvements over the Newcomen-type engine. Watt still used a vacuum but he replaced the atmospheric air pressure by steam pressure, which

Figure 1-6. Boulton & Watt pumping engine, c. 1777, for the
Birmingham Canal Navigations. *(Courtesy of The Science
Museum, London)*

kept the piston and cylinder hotter and improved efficiency (see Fig.
1-9). In the next year, 1776, a third engine was built and over the next
15 years Watt made many improvements.

The refinements were evaluated by comparing how many bushels
of coal were burned with how much water was pumped. The less coal
burned per gallon of water pumped, the more efficient the engine.

Calculating the size of an engine to pump water was relatively
simple, but when horses were being replaced by machines the problem
was more complex. In 1782, while studying the problem of replacing
"horsepower," he observed the fact that an engine had to have the
equivalent power of a certain number of horses to operate a mine
pump. His calculations indicated that one horse could raise 33,000
pounds one foot in one minute and this figure is still used today in
measuring "horsepower."

The partnership of Watt and Boulton continued to yield improve-

Figure 1-7. Matthew Boulton, 1801. *(Courtesy of The Science Museum, London)*

ments to their engines. The use of steam as a motive power was well established by the early 1800's, and many inventors and engineers made additional refinements. The introduction of mechanical cranks and sun-and-planet gearing (see Fig. 1-10) changed the vertical motion to rotative. This gave the steam engine the ability to drive a wheel which, in turn, powered factory machinery, paddle wheels on ships, and railway engines.

The Industrial Revolution had begun about 1750, and created a vast demand for power. The work of the dedicated and resourceful men who harnessed the power of steam gave the Industrial Revolution the power it needed. As we shall see in later chapters, the intelligent and efficient use of steam was aided by many men. The reader engaged in the daily production or use of steam will find himself part of a long tradition extending back to Savery, Newcomen, Boulton, and Watt.

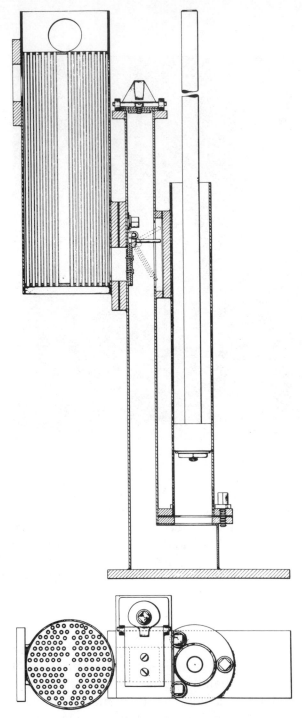

Figure 1-8. Watt's tubular surface condenser, c. 1765.
(Courtesy of British Crown Copyright, The Science Museum, London)

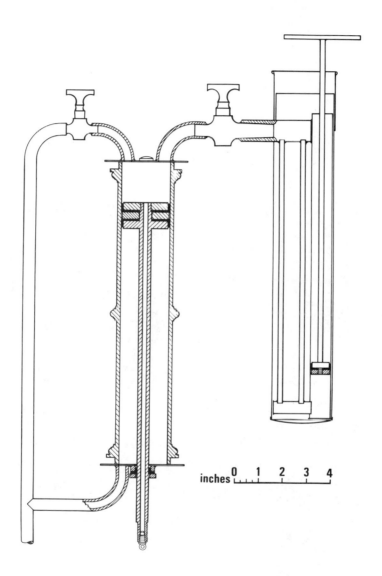

Figure 1-9. Watt's first model (conjectural restoration). *(Courtesy of The Science Museum, London)*

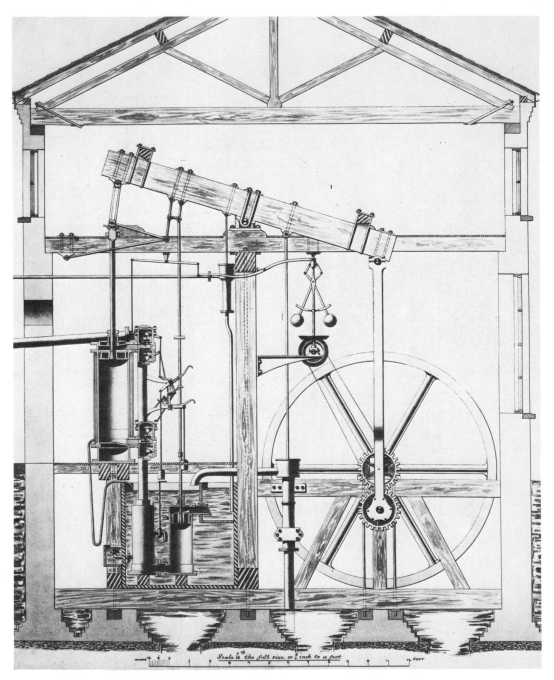

Figure 1-10. James Watt's double acting rotative beam engine
with sun and planet gearing, c. 1788. *(Courtesy of British
Crown Copyright, The Science Museum, London)*

MULTIPLE CHOICE QUESTIONS

1. Hero's Aeolipile used the principle of the:
   a. Reaction turbine
   b. Windmill
   c. Lever

2. Giovanni della Porta described experiments using steam in:
   a. 1750
   b. Early 18th century
   c. 1606

3. A Frenchman living in England devised an ingenious use for steam. It was a:
   a. Locomotive
   b. Fountain
   c. Carousel

4. Galileo realized that water could not be pumped from a depth greater than:
   a. 28 ft
   b. 50 ft
   c. 32 ft

5. The principle of the barometer was first demonstrated by:
   a. James Watt
   b. Torricelli
   c. Galileo

6. Early experiments concerning atmospheric pressure and vacuum were conducted in:
   a. Germany
   b. France
   c. England

7. Robert Boyle made significant contributions to the science of:
   a. Chemistry
   b. Atmospherics
   c. Thermodynamics

8. The system of rods and levers on a beam engine was known as:
   a. Rod levers
   b. Beam cords
   c. Potter cords

9. James Watt was:
   a. An engineer
   b. An instrument maker
   c. A scientist

10. Watt's first patent on the atmospheric engine was issued in:
    a. 1750
    b. 1800
    c. 1769

2

Temperature and the Effects of Heat

Before man discovered the use he could make of fire he was no doubt aware of the difference between the conditions of heat and cold. Man first used fire to warm himself and cook his food and later to smelt metals, make glass, and even burn witches and heretics at the stake. He was aware of what heat and cold could do, but for thousands of years had no way of measuring heat and cold, as we do by its temperature. The temperature of a body of matter is a measure of the tendency of that body to transmit heat to, or withdraw it from, another body.

TEMPERATURE MEASUREMENT

Modern techniques for heat measurement date back to the 18th century. In 1714, in Danzig, Germany, Gabriel Daniel Fahrenheit assumed that a mixture of ammonium chloride and snow in equal parts was the lowest temperature possible. He fixed this at zero degrees. The blood temperature of a human body as shown by the rise in a column of mercury was chosen as marking another distinct point in this scale and was called 100 degrees. The distance between these two points was divided into 100 equal parts, each called a degree of temperature. Fahrenheit was wrong in his conclusion, and as we know today blood heat is 98.6°F, and the lowest possible temperature is far below his zero mark. The accepted temperatures today are the freezing and boiling points of water, measured on the Fahrenheit scale as 32 °F and 212 °F.

In 1742 a Swedish astronomer and physicist, Anders Celsius, did take the freezing and boiling points of water as the definitive limits of temperature and called them zero degrees and 100 degrees. This thermometer is known as the Celsius or centigrade scale.

A third system in use is the Reaumer scale, on which thermometers read from 0° to 80°. This scale is used principally in the Soviet Union and other parts of Eastern Europe.

Fahrenheit was off the mark when he determined his zero degree minimum. The *absolute* thermometer scale must be used in certain measurements when dealing with some substances, especially gases. When a perfect gas, confined in a vessel at 0° on the absolute scale is cooled 1 °C, its pressure will decrease 1/273 of the pressure at 0 °C. From this observation, if the temperature were lowered 273 °C below 0 °C there would be an absence of all heat. The temperature at this point would be 0 °C on the absolute scale and water would freeze at 273 °C absolute and boil at 373° absolute. On the Fahrenheit scale, absolute zero would be 492 °F below the freezing point of water which is 32 °F; therefore, 0 °F is the same as 460° absolute. With both the Fahrenheit and centigrade scales in common use, it is often necessary to convert from one to the other. In placing Fahrenheit and centigrade thermometers side by side we see that between 0° and 100° on the centigrade scale there are 100 divisions; on the Fahrenheit scale between the same points, that is, the boiling point and freezing point of water, the division is 212° − 32° = 180°. Therefore, for every degree centigrade there are 1.8° Fahrenheit (see Fig. 2-1):

$$1 °C = 1.8 °F \quad \text{or} \quad \frac{9°}{5} F$$

and

$$1 °F = 0.555 °C \quad \text{or} \quad \frac{5°}{9} C$$

To convert temperatures shown on the centigrade scale to Fahrenheit, multiply the number of degrees centigrade by $\frac{9}{5}$ and add 32.

The formula is:

$$F° = \frac{9}{5} \times °C + 32$$

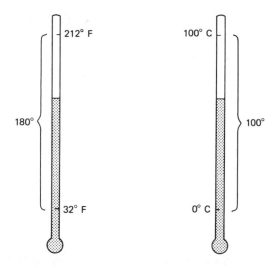

Figure 2-1. Comparison of Fahrenheit vs. centigrade.

Example:

A thermometer reads 10 °C. What is this reading in degrees F?

$$F° = \frac{9}{5} \times °C + 32$$

$$= \frac{9}{5} \times 10 + 32$$

$$= 9 \times 2 + 32$$

$$= 50°$$

To convert temperatures shown on the Fahrenheit scale to centigrade find the number of degrees above the freezing point of water and multiply by 0.555 or $\frac{5}{9}$. The formula is:

$$C° = \frac{5}{9}(F° - 32)$$

Example:

A quantity of water is heated to 82 °F. What is this temperature in degrees C?

$$C° = \frac{5}{9}(82 - 32)$$

$$= \frac{5}{9} \times 50$$

$$= \frac{250}{9} = 27.7°$$

To convert a thermometer reading to the absolute scale:

For centigrade add 273 °C.

For Fahrenheit add 460 °F.

By formula:

$$C° \text{ absolute} = C° + 273$$
$$F° \text{ absolute} = F° + 460$$

HEAT

All of the earth's heat is derived directly or indirectly from the sun. This solar heat promotes plant life, which eventually is transformed into wood or coal. Coal is formed from deposits of vegetable matter that have been subjected to heat and pressure over millions of years. The decay of animal and vegetable matter within the earth results in deposits of oil and natural gas. The solar heat is stored in these fossil fuels and is thus indirectly made available for man's use.

Heat is also produced by chemical reactions. When certain chemical processes or reactions occur, energy that resided in the substances involved in the reaction are liberated as heat energy.

Chemical reaction includes all forms of combustion, and combustion or burning of fuels is the source of heat with which a power plant operator is concerned. The burning of coal, wood, oil, or gas is the chemical action between the carbon and hydrogen contained in the fuel and oxygen in the air, thus producing heat.

The amount of heat in any substance is determined by:

    (a) the quantity of the substance
    (b) the nature of the substance
    (c) whether it is solid, liquid, or gaseous
    (d) its temperature.

Therefore, if we know all the above factors we can determine how much heat is available.

To find out the quantity of heat in a given body, it is necessary to use a measuring unit. The unit used to measure heat is called a British thermal unit (Btu). A British thermal unit is defined as the amount of heat required to raise the temperature of one pound of water 1 °F. Using the above definition, it is possible to determine accurately the amount of heat required to raise steam in a boiler or heat a tank of water.

To evolve this into a formula, we can call Q the quantity of heat required, W the weight of the water, and T_2 and T_1 the desired and original temperature of the water. Thus

$$Q = W(T_2 - T_1)$$

Example:

How much heat in Btu's is required to raise the temperature of 100 gallons of water from 60 °F to 100 °F. Assume that each gallon weighs eight pounds.

$$
\begin{aligned}
Q &= W(T_2 - T_1) \\
&= 100 \times 8(100 - 60) \\
&= 800(40) \\
&= 32{,}000 \text{ Btu's}
\end{aligned}
$$

As we know, it is possible to convert the heat energy in a fuel into steam and use that steam to drive a pump or engine. Therefore, we are converting heat energy into mechanical energy. The relationship between other forms of energy and heat is the subject of thermodynamics. The first law of thermodynamics states that heat and mechanical energy can be converted from one to the other, and therefore a definite relationship exists between them. Since such a relationship exists, it follows that there must be a mechanical equivalent of a Btu.

The foot-pound is the unit of mechanical energy first proposed by James Watt, and a Btu is equivalent to a verifiable number of foot-pounds. For practical purposes, one Btu is equal to 778 ft-lb. One horsepower is equivalent to 33,000 lb being moved one foot in one minute; therefore,

$$1 \text{ hp} = 42.42 \text{ Btu's per minute}$$
$$1 \text{ hp} = 2545 \text{ Btu's per hour}$$

The second law of thermodynamics states that heat has never been known to flow from a cold body to a hotter body of its own accord. This means that the flow of heat is always from a hotter body to a colder body; hence, to cause a reverse condition as in a refrigeration machine, an expenditure of energy is required.

British Thermal Unit

A Btu is defined as the amount of heat required to raise one pound of water 1 °F; however, the effect of heat on one pound of other substances should be considered. Every substance can be raised in temperature 1 °F, but requires a varying number of Btu's to do so. This amount of heat is called the *specific heat* of the substance. The specific heat of water is 1.000, and is the greatest specific heat of any substance.

Specific Heat

Specific heat is defined as the number of Btu's required to raise the temperature of a substance 1 °F. That various materials have different specific heats can readily be verified by taking a one-pound ball of iron and a one-pound ball of copper and heating both to the same temperature. Drop each ball into the same amount of water at the same temperature, and it will be seen that the water into which the iron ball was dropped attains the higher temperature. The specific heat of iron is 0.110 and the specific heat of copper is 0.093. This shows that a pound of iron at a given temperature contains more heat than the same amount of copper at the same temperature.

Latent Heat

The heat energy created in a boiler is primarily used to make steam, and to do this, water is vaporized. When a liquid is vaporized, there is no change in temperature, but a substantial amount of heat is consumed in doing the work of vaporization.

The heat energy so used is called *latent heat* and is that energy required to produce a change in the physical state of a substance whether the change be from solid to liquid, as when ice melts, or from liquid to a vapor, as in the production of steam. The reverse process, as in the condensation of steam and the freezing of water, involves the extraction of heat from the substance.

When ice or a metal melts, the heat utilized is called the *latent heat of melting.* When we reverse the process and solidify a liquid, the heat extracted is called the *latent heat of fusion.*

The heat energy required to vaporize a liquid is called the *latent heat of vaporization,* and when we reverse the process, the heat extracted is called the *latent heat of condensation.*

The *total heat* of a body is a combination of the observed sensible heat and the latent heat added (if the body has changed state).

To convert the heat energy in a pile of coal or a tank full of oil to steam, the heat energy must be transferred to the water in the boiler. Heat is transferred in three ways:

        (1) by conduction
        (2) by convection
        (3) by radiation.

Heat transfer can occur simultaneously by all three methods.

Conduction

If a copper bar is held in the hand at one end and the other end is heated, after a while the heat being applied at the one end will be felt at the other end of the bar. The transfer of heat from one end of the bar to the other is by *conductance.* The heat you feel is *conducted heat.*

The factors that affect the transfer of heat by conductance are:

        (1) the nature of the substance
        (2) the cross-sectional area
        (3) the temperature difference between one
            end of a body and the other
        (4) the length of time that heat flows
        (5) the length of the body or path through the
            substance.

Different substances have different rates of conductance. Metals are very good conductors, while wood is a poor conductor, and we utilize fiberglass and asbestos as insulation material because they conduct heat so poorly. When specifying insulating materials, the guide as to which is most suitable is governed by its K factor, which is a measure of its resistance to flow of heat.

Convection

Conductance takes place in a boiler when the metal of the fire box or the boiler become heated and the heat is transferred to the water in the boiler. When the heat has passed through the metal to the water, another form of heat transfer occurs. This is *convection,* or the transfer of heat by the flow of currents in a fluid. This flow is caused by a change in the density of the fluid. It is well known that hot air rises and so does hot water. As the water becomes heated the hotter water rises to the surface, displacing the colder water, which is heated in turn. Continuous circulating currents are set up in the fluid.

The transfer of heat by convection is utilized in many ways. Hot-water heating systems depend upon this principle for their effectiveness. Boilers are designed to take advantage of the convection currents to achieve maximum heat transfer.

Radiation

A third means of transferring heat is by radiation. If you hold your hand near a hot pipe that is not insulated you will feel the heat of the pipe, without touching the pipe. The transfer of heat has taken place through the air and is called *radiant heat*. Radiant heat is felt because it travels in waves and is not conducted or convected. Heat can be radiated through a vacuum and, in fact, the heat from the sun travels millions of miles through the vacuum of space.

A boiler receives radiant heat from the gases of combustion. It will also receive some radiant heat from a luminous flame, but most of the radiant heat transferred to the boiler water comes from the firebrick lining and, in the case of coal fires, from the glowing fuel on the fire grates.

It is to the operator's advantage to maintain as high a furnace temperature as possible to achieve the maximum transfer of radiant heat energy. In a steaming boiler, if the fire is shut down, the boiler will continue to produce steam for some time, utilizing the radiant heat available from the glowing firebricks.

After understanding the various ways in which heat is observed, how it is measured and how it flows, we must now consider the effects of that heat as it applies to boiler operation.

EFFECTS OF HEAT

The addition of heat to any solid causes it to expand and when the temperature is decreased it contracts. The one exception is water. When water is cooled to 39.2 °F (4 °C), it contracts. Below this temperature it expands and, from everyday experience, we know what ice can do to water pipes, when the water has frozen inside them. One cubic foot of water becomes 1.085 ft³ of ice. As we have seen, different substances have different specific heats, which causes them to absorb varying quantities of heat; so likewise the expansion and contraction of solids varies among different substances at the same temperature.

Expansion and contraction take place in three ways:

(1) lineal
(2) areal
(3) volumetric.

Each substance can expand and contract a certain amount per Fahrenheit degree. The amount of increase or decrease in the length of a substance when heated or cooled is called the *coefficient of linear expansion.*

Areal expansion is the increase in the area of a surface when heated and the *coefficient of areal expansion* is taken as twice the coefficient of linear expansion. This figure is not precise, but for practical work, the error is negligible.

Volumetric expansion is the expansion of a volumetric figure such as a cube, and is of value in boiler operations because all liquids such as water and oil expand volumetrically. *Coefficients of volumetric expansion* of liquids are difficult to determine because the liquid is contained in a vessel made of a material that has a different coefficient of expansion than the liquid.

Liquids are for all practical purposes incompressible, and if we fill a container with a liquid and heat it, the pressures that can be achieved are enormous. When a hot-water heating system is filled with cold water and then heated, the pressure rises considerably. The addition of an expansion tank to these systems helps to relieve these pressures, without the use of a safety valve that would allow the system to become depleted in water content.

When water at a temperature above 39°F is heated, it expands, its density decreases, and it becomes "lighter"—that is, its weight per unit volume becomes less. In a boiler this condition makes possible the convection currents previously discussed.

EFFECT OF HEAT ON GASES

All operating engineers are aware of gases in the boiler room. These gases may be the ones that are burned for fuel or they may be the products of combustion which go through the boiler and up the stack. All of these gases are affected in some way by heat, and it is of practical importance, because of their application to the various heat processes of which the engineer must deal, that he be aware of the principles involved in dealing with such gases.

Before we proceed, it is necessary that we understand what a gas is. The definition of a gas is any substance in the vapor state which closely follows the general gas law. When dealing with gas laws, we are concerned with Robert Boyle's work. Boyle's law, named after the Irish physicist, states that if the temperature of a gas is kept constant, the absolute pressure of the gas will vary inversely as its volume; conversely, the volume will vary inversely as the absolute pressure.

The formula for Boyle's law, which follows directly from the previous statement is:

$$\frac{P_1}{P_2} = \frac{V_2}{V_1}$$

where P_1 and P_2 are the initial and final absolute pressures of any weight of gas at a constant temperature, and V_1 and V_2 are the initial and final volumes of a gas, both volume units being the same. Graphically, the formula would be as shown in Fig. 2-2.

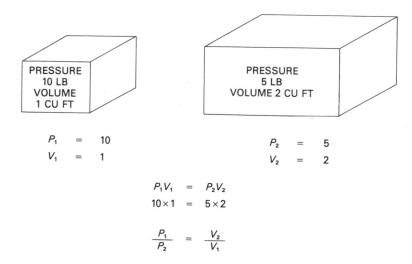

Example

A pressure gauge on a cylinder of 4-ft³ volume reads 20 PSIA. A valve is opened and all the gas is transferred to a cylinder of 8-ft³ volume. What is the pressure of the gas?

$$\frac{P_1}{P_2} = \frac{V_2}{V_1}$$

$$\frac{20}{P_2} = \frac{8}{4} = \frac{20 \times 4}{8} = \frac{80}{8} = 10 \text{ PSIA}$$

It must be understood that this formula will apply where the volume of a given weight of gas at a certain pressure is known and it is desired to find out the volume of the gas at a different pressure. Gas temperature will remain the same for both conditions. The transposed forms of Boyle's law which are useful in solving problems involving pressure/volume of a gas are:

$$P_2 = P_1 \frac{V_1}{V_2}$$

$$V_2 = P_1 \frac{V_1}{P_2}$$

All of the symbols have the same meanings as previously stated.

The previous law deals only with changes in pressure and volume and assumes a constant temperature. The effect of a temperature change on a gas was investigated by Jacques Charles and Joseph Gay-Lussac early in the 19th century. The results of their experiments indicated that gases follow much the same general law of expansion as solids and liquids.

Charles's law may be stated thus: If the volume of a given weight of gas is kept constant, the absolute pressure will vary directly as the absolute temperature of the gas.

The problems to which Charles's law may be applied are those in which the pressure of a given weight of gas, which is maintained at a constant volume, is known for a certain temperature, and it is desired to know either the pressure exerted by the gas at a different temperature or the temperature necessary to produce a different pressure. The formula as derived is:

$$\frac{P_1}{P_2} = \frac{T_1}{T_2}$$

By transposing:

$$P_2 = \frac{P_1 T_2}{T_1}$$

$$T_2 = \frac{T_1 P_2}{P_1}$$

The equations for Boyle's law and Charles's law may be combined to provide us with a formula which will deal with the pressure, volume, and temperature of any gas. If we have a definite amount of gas of known composition, the condition of that gas is completely specified if we know any two of the quantities. The combined formula is:

$$\frac{P_1 V_1}{T_1} = \frac{P_2 V_2}{T_2}$$

and by transposing:

$$P_2 = \frac{P_1 V_1 T_2}{V_2 T_1}$$

$$V_2 = \frac{P_1 V_1 T_2}{P_2 T_1}$$

$$T_2 = \frac{P_2 V_2 T_1}{P_1 V_1}$$

The majority of gases can be liquefied if subjected to sufficient pressurization. When they revert to their gaseous state, they vaporize. The phenomenon of vaporization is important to the boiler operator because the production of steam is the vaporization of water. There are two kinds of vaporization: the first is *evaporation* and the second *ebullition.*

Evaporation only occurs on the surface of a liquid. Ebullition or boiling, on the other hand, takes place in the body of the liquid, and the vapors pass through the liquid to the surface. The factors governing the rate of evaporation are:

(1) it increases with temperature
(2) it is greater into dry air than into wet air
(3) it increases if you decrease the pressure at the interface with the forces tending to vaporize.

Condensation

Condensation is the opposite of evaporation, that is, it is the process by which a substance changes from a gaseous to the liquid state. Condensation is produced by taking heat away from the vapor. So long as the pressure remains constant the vapor will condense. In a steam power plant, condensers are utilized to liquefy the exhaust steam from engines and from cooling coils in order to reuse the condensate as boiler feed water.

When water is vaporized it will increase in volume. For example, if we have one cubic foot of water at 212°F, which is the boiling temperature of water, it will expand to about 1600 ft³ of steam at atmospheric pressure.

All gases have what is known as vapor pressure, which is the actual pressure exerted by the body of a vapor. Vapor pressure is expressed in pounds per square inch or any other measure of pressure. The vapor pressure that we refer to is actually the steam pressure in the boiler.

As mentioned earlier, evaporation will increase if the air is dry. If the air is not dry, the vapors of evaporation will be saturated. Saturated vapor is any vapor from which heat cannot be extracted without partially condensing. The steam in a boiler is a saturated vapor and only by adding heat, such as through superheating tubes, can we change that condition. By raising the temperature without increasing the pressure, we create what is known as superheated steam.

The ebullition or boiling of a liquid generates heat within the body of the liquid and, as in the boiler, the liquid in contact with the shell of the boiler absorbs the heat from the fire. The steam bubbles that are formed at that point rise to the surface, giving up their heat as they go, and setting up the process of convection. The convection process continues until the entire mass of liquid acquires a uniform temperature. The bubbles which are formed at the bottom of the vessel retain their heat and rise to the surface where they break and their vapor will be mingled with the vapor or steam already generated.

Steam may occur in any one of three conditions: saturated steam, superheated steam, or supersaturated steam.

Supersaturated steam is steam that is cooled by its own expansion until it contains less heat energy than saturated steam under the same conditions. This condition is very unstable, and the steam soon resumes its saturated condition. Supersaturated steam is not of importance in ordinary power plant practice and is rarely considered except in the design and operation of turbines. It is mentioned here merely to acknowledge its existence.

The usual condition of steam in power plant operations is either saturated or superheated. Saturated steam is steam at the temperature corresponding to the boiling point of the liquid at the pressure imposed on the liquid. There is, for each different pressure, a definite boiling point.

When heat is added to a confined liquid, as in a boiler, and the pressure remains constant, more of the liquid vaporizes but the temperature of the liquid and the vapor remain constant. The temperature will not change until all the liquid is vaporized, because the heat we are adding is the latent heat of vaporization. For a given pressure, the temperature of saturated steam is fixed and we cannot increase that temperature without raising the pressure. Saturated steam may be wet or dry. A dry saturated steam is one that does not contain any liquid. A wet saturated steam contains liquid particles in suspension.

Superheated steam occurs at a temperature greater than the boiling point corresponding to the pressure imposed on it. The temperature of the steam may be increased by adding heat to it after all the liquid has been vaporized or the steam has been removed from contact with the liquid. The amount of superheat is the difference between the superheated temperature and that of a saturated steam at the same pressure. Steam which has a temperature 30° higher than saturated steam at the same pressure is said to contain "30° of superheat."

Steam has various properties and these properties are interrelated, so that if some of them are observed others may be found. Pressure gauges, thermometers, and steam calorimeters are used to determine the properties of steam at various points. A steam calorimeter is a device for measuring the amount of liquid in the steam. However, if it is known that the steam is dry and saturated, only one other property such as pressure need be known; the knowledge that it is dry constitutes the second known property which is "quality."

The quality of saturated steam is a measure of the percentage of dry steam present in a given amount of wet saturated steam. If, for instance, 100 lb of wet steam contain two pounds of liquid, it must contain 98 lb of dry steam. That steam would then be characterized as steam with a dryness fraction of 0.98.

The properties of steam, used in general practice, are arranged in what are known as "steam tables." Table 2-1 gives values for dry

TABLE 2-1
Steam Table

ABS. P. LB/IN.^{2}	GAUGE LB/IN.^{2}	TEMPERATURE F		HEAT IN BRITISH THERMAL UNITS PER LB				
		t	T	st$_o$	G	h	L	H
15	0.3	213.0	672.7	180.4	29.8	181.2	970.0	1151.2
20	5.3	228.0	687.7	195.3	34.6	196.3	960.4	1156.7
25	10.3	240.1	699.7	207.4	38.8	208.6	952.5	1161.1
30	15.3	250.3	710.0	217.6	42.4	219.0	945.6	1164.6
35	20.3	259.3	719.0	226.6	45.8	228.0	939.6	1167.6
40	25.3	267.2	726.9	234.4	48.8	236.1	934.4	1170.5
50	35.3	381.0	740.7	248.2	54.4	250.2	924.6	1174.8
60	45.3	292.7	752.4	259.8	59.2	262.2	916.2	1178.4
70	55.3	302.9	762.6	270.0	63.8	272.7	908.7	1181.4
80	65.3	312.0	771.7	279.0	67.6	282.1	901.9	1184.0
90	75.3	320.3	780.0	287.3	71.5	290.7	895.5	1186.2
100	85.3	327.8	787.5	294.8	74.9	298.5	889.7	1188.2

saturated steam as determined accurately by experiment. The heat content values shown in the steam table gives the Btu content of one pound of steam above 32 °F liquid.

There are two tables—one gives the properties of dry saturated steam and the other the properties of superheated steam.

The heat of the liquid, *h,* as given in Table 2-1, is the heat required to raise one pound of water from 32° to that pressure at which the water begins to boil at the pressure *P.* In changing water to steam, under constant pressure, the temperature of the water must be brought to its boiling point at the given pressure before the water can evaporate. The heat required to do this is called the heat of the liquid.

The latent heat of vaporization, *L,* is also given in the table. The latent heat of vaporization is the heat required to convert the water into steam.

The total heat of the steam *H* is the amount of heat which must be supplied to the water at 32°F, to convert it, at constant pressure, into steam at the pressure and temperature desired. The total heat is the sum of the heat of the liquid and the latent heat of vaporization.

The total heat of steam depends upon how it has been heated. The table gives the values for heating at constant pressure, as when a boiler is already under steam. To find heat values at different pressures and temperatures, the amount of heat can be calculated by using the table.

Example:

Find the amount of heat which must be supplied to 53 lb of water at 59 °F to convert it into steam at 250.3 °F.

The heat of the liquid at 250.3°F is 219.0 Btu's per lb. At 59°F it is 27.00 Btu's/lb. The amount of heat that must be added is

$$219.0 - 27.00 = 192.00 \text{ Btu's/lb}$$

The heat required to vaporize the water at 250.3 °F is 945.6 Btu's per pound. Therefore, the total heat required is

$$(945.6 + 192.00) \times 53 = 60,292.8 \text{ Btu's}$$

The total heat content of the 53 lb of water converted to steam is 60,292.8 Btu's.

We must now introduce the idea of *enthalpy*. This is a word of Greek derivation which is used when describing the heat content of steam. Instead of saying the heat content is so many Btu's per pound we say that the enthalpy of the steam is a certain figure, although this is another way of saying the same thing. A dictionary definition of enthalpy is "the quantity of heat in a substance or physical system per unit of mass." To be technically correct, when expressing the heat content of steam or liquid, we should call it the enthalpy of the liquid or the enthalpy of the steam, etc.

In later chapters we shall make further use of the steam tables in the calculation of boiler efficiency.

EXERCISES

1. Define "latent heat."
2. Define "sensible heat."
3. State the first and second laws of thermodynamics.
4. Who was Robert Boyle?

5. How many degrees Fahrenheit are in 1 degree Celsius?

Ans.:   1.8

6. The temperature reading on a feed heater is 235 °F. Convert this reading to centigrade.

Ans.:   112.7 °C

7. You are told that the temperature of a liquid is 480 °F absolute. Could this liquid be water?

Ans.:   No, water freezes at 32 °F

8. A boiler contains 1000 gallons of water. Temperature of the water is 60 °F. How much heat in Btu's must be added to the water to raise its temperature to 212 °F at atmospheric pressure? Assume water weighs 8 lb/gal.

Ans.:   1,216,000 Btu's

9. In Question 8, if the water is heated in 1 hour's time, how many horsepower has the water absorbed?

Ans.:   477.79 hp

10. If the specific heat of a certain material is known to be .13 how many Btu's will 1 lb of the material absorb in order to raise its temperature 10 °F.

Ans.:   1.3 Btu's

11. A cylinder contains a gas at an absolute pressure of 250 lb/in². The internal volume of the cylinder is 4 ft³. If the gas is released into another closed container with a volume of 10 ft³, what will be the absolute pressure in the second container?

Ans.:   100 lb

12. A tank containing a gas at an absolute pressure of 200 lb/in² and a temperature of 50 °F is exposed to the sun. The temperature rises to 75 °F. What is the internal pressure at the increased temperature?

Ans.:   209.8 lb/in²

13. If a quantity of steam is said to have a dryness fraction of 0.95, how much water, in pounds, does 1000 lb of such steam contain?

Ans.:   50 lb

MULTIPLE CHOICE QUESTIONS

1. Absolute zero on the centigrade scale is:
   a. 273°
   b. 492°
   c. 0°

2. Each centrigrade degree is equal to:
   a. 1.8°F
   b. $\frac{5}{9}$°F
   c. 1°F

3. One Btu is equal to:
   a. 778 ft-lb
   b. 1 hp
   c. 1°F

4. One horsepower is equal to:
   a. 2778 Btu/hr
   b. 2454 Btu/hr
   c. 2545 Btu/hr

5. The specific heat of copper is _____ than iron.
   a. more
   b. less
   c. same as

6. The type of heat that changes the physical state of a substance is:
   a. Specific heat
   b. Sensible heat
   c. Latent heat

7. The movement that water achieves when heated is called:
   a. Radiation currents
   b. Conduction currents
   c. Convection currents

8. The volume of a gas will vary _____ as the absolute pressure.
   a. conversely
   b. inversely
   c. directly

9. One cubic foot of water evaporated to steam will produce _____ cubic feet of steam at atmospheric pressure.

a. 1600
b. 16,000
c. 160

10. The enthalpy of steam is the Greek word for:
a. Temperature
b. Heat content
c. Specific gravity

3
Fuels

Ultimately, all fuels derive their energy content from the sun, but the selection of a fuel for power plant operation or any other use is governed by the proximity of the fuel to the point of its use as well as its cost and heating value.

Manufacturers of wood products have a readily available source of fuel in their factory scrap. The crushed cane on a sugar plantation, called bagasse, is an economical fuel for a crushing mill, and straw residue from the threshing of grain provides a fuel in agricultural regions.

All fuels like other matter, are divided into three general categories: solid, liquid, and gaseous. The most important and widely used solid fuel available in quantity to a power plant is coal. Its abundance, ease of handling and heating value were responsible for its popularity as the fuel to power industry and transportation from the mid-18th century to modern times.

All solid fuels are organic in origin. Coal was formed from deposits of organic matter during the Carboniferous Period of the earth's history, about 300 million years ago.

COAL

Coal is a complex substance consisting of many variations of chemical compounds, but all coals contain the basic components of hydrogen and carbon, the elements that contribute the largest amount of heating value to coal.

The types of coal mined in the United States are classified as anthracite, bituminous, semianthracite and subbituminous, depending upon the amount of organic volatile matter contained in each type.

Anthracite is practically all carbon. It burns slowly with little smoke. Nearly all anthracite used in the United States comes from Pennsylvania. It is usually called hard coal and has a heating value of 13,000 Btu's per pound of dry coal. Since it contains little volatile matter, it burns with a very small flame. Anthracite has been the most commonly burned coal for heating residences and is available in several sizes for this use. The common names for types of anthracite are pea, nut, stove, and egg. The sizes range from pea, with a maximum diameter of ⅞ in., to egg at 3¼ in. diameter. Semianthracite burns more rapidly and with a longer flame than anthracite because of its higher content of volatile matter. It produces less slag and little smoke.

Bituminous coals are by far the most commonly used by industry. Because of variations in their percentages of volatile matter, some bituminous coals burn freely with a short flame while others have a longer flame. Heating values of bituminous coal vary between 11,000 to 14,000 Btu's per pound.

Coal is sold at varying prices, depending upon quality, size, and other factors, but the important considerations for a power plant operator are its Btu content and chemical composition.

There are two kinds of analysis for coal: the proximate analysis and the ultimate analysis. The proximate analysis determines the content in percentages of four components—moisture, volatile matter, fixed carbon and ash. The proximate method is designed for quickly establishing the qualities in the coal important in the operation of a boiler, and it is a general practice to consider along with the analysis the heating value of the fuel. The tests in an analysis simulate the actions that occur when the coal is burned in a boiler, and in addition show the losses in the weight of a sample due to moisture.

To make a proximate analysis, we use the following procedure. Take a representative sample from a coal pile and crush it into a powdered form that will pass through a 20-mesh sieve. Place the powder, usually 5 to 10 lb, in an airtight container. Remove 2 to 3 oz from the container and set them aside, resealing the container to prevent moisture from entering.

Weigh 4 grams (approximately ⅛ ounces) of the powder in a platinum crucible. The crucible should be weighed first before you place the powder in it. This should be done as quickly as possible to prevent a loss of moisture into the atmosphere. Note the combined weight of the crucible and coal. Place the crucible in an oven and heat it to 230°F for one hour. Remove the crucible from the oven and allow the crucible to cool in a dry box or dessicator. When cool, weigh the crucible holding the sample and note the amount of loss in weight. Place the crucible in the oven again and heat for a half hour. Again cool the crucible and weigh it. If there is no further reduction in weight all moisture has been removed from the coal, and the difference between the first and last readings indicates the moisture content.

Now weigh a new sample, about 1 gram, also in a platinum crucible with a tight-fitting cover. Place the crucible on a Bunsen burner and heat for 7 to 8 minutes. Cool the crucible as before and weigh it. The loss in weight is the sum of the volatile matter plus the moisture.

Remove the cover from the crucible that has just been used for the "volatile" test and heat the crucible over a Bunsen burner until all the carbon has burned off. Stir the contents of the crucible, which should be a powdery mass when combustion is completed. Cool as previously and weigh the crucible and its contents. The difference between this weight and the weight of the empty crucible is the weight of the ash. The weight of the carbon is determined by subtracting the sum of the weights of the mositure, ash, and volatile matter from the original weight of the sample.

The percentage of each constituent should give you a basis to judge the heating value and general suitability of the coal for power plant operation.

If the ash content is high, it can be assumed that the coal is generally of poor quality. The amount of volatility indicates whether the coal will burn with a short or long flame and whether it will tend to produce smoke. The more volatile the coal, the more it will smoke and require more excess air to comply with Federal Environmental Protection Agency regulations. The combustibles in any coal are the volatile matter and fixed carbon; generally a high percentage of these constituents indicates a coal of high heating value.

The ultimate analysis of coal is made to determine the percentage of the weight of all the chemical constituents of the coal. These are carbon, hydrogen, oxygen, nitrogen, and sulfur and the amount of the ash remaining after the test. This analysis is performed by an analytical chemist and a description of the process is not necessary here. After an ultimate analysis has been made, the heating value of the fuel is determined by substituting in the formula derived by Pierre Louis Dulong (1785–1838), a physicist and chemist who was a professor and director of the École Polytechnique in Paris. His experiments to determine the specific heat of elements added immensely to knowledge of the subject. Dulong's formula is:

$$H = 14{,}500\,C + 62{,}000\,(H - \frac{O}{8}) + 4000\,S$$

H is the heating value of coal in Btu's/lb and C, H, O, and S are the weights of the carbon, hydrogen, oxygen, and sulfur contained in one pound of coal. The weights of the constituents contained in one pound of the fuel may be determined by dividing the percentage values of the ultimate analysis by 100. If the ultimate analysis shows 86.68% of carbon, divided by 100, there is 0.8668 pounds of carbon per pound of coal.

Suppose an ultimate analysis of a type of coal shows the following percentages of constituents:

Carbon	88.86
Hydrogen	2.04
Oxygen	1.95
Sulfur	0.35

The heating value of the coal is:

$$H = 14{,}500\,C + 62{,}000\,(H - \frac{O}{8})\ 4000\,S$$

$$= (14{,}500 \times 0.8886) + 62{,}000\,(0.0204 - \frac{0.0195}{8}) + (4000 \times 0.0035)$$

$$= 12{,}884.7 + 1113.675 + 14$$

$$= 14{,}012.375\ \text{Btu's/lb}$$

Another method to determine the heating value of a fuel involves a device known as a fuel calorimeter. Several types are manufactured, but all involve the same principle: that is, the burning of a fuel in the presence of a known weight of water, measuring the rise in the water temperature and then converting this into Btu's released by the fuel when combustion is completed.

The accuracy of a fuel calorimeter depends on two factors: the completeness of combustion and the amount of heat lost. Various mechanical arrangements have been devised so that inaccuracy due to these factors has been reduced to a minimum.

The selection of the proper fuel for all energy conversion units is an important economic one and in the present energy situation requires thorough understanding of all factors involved. The most important items to consider are:

(1) the kind or kinds of fuel available nearby
(2) the cost of various fuels
(3) their heating value
(4) the cost of handling
(5) prospective availability of supply
(6) cleanliness
(7) plant requirements
(8) capital costs
(9) depreciation of capital investment.

OIL

The supply factors and economic developments of the last 30 years have been responsible for large-scale conversion from burning of coal to oil and natural gas, but the diminishing availability of domestic oil and gas is causing a return to coal as the major fuel in many installations. For the time being, however, it appears likely that a massive amount of fuel oil will continue to be burned.

Man has been using oil products for more than 4000 years, but the birth of the vast modern petroleum industry occurred in Titusville, Pa., on August 27, 1859 when Colonel Drake, in drilling for oil, found it at a depth of 69 feet.

Since that day, the use of petroleum products has risen phenomenally, and with the scientific advances in the last century oil has become of vital importance to our economy and living habits.

Geologists tell us that petroleum comes from the decomposed remains of marine animals and plants that existed millions of years ago. The warm seas covering most of the earth's surface contained an abundance of plant and animal life. Rivers flowed into these seas, carrying mud and sediment that was deposited over the remains of plant and animal life. Sealed from air, the plants and animals slowly decomposed and were transformed into petroleum and gas. These are the sources of the products that we use today in refined form to power automobiles and fuel boilers.

Crude oils are thick liquids that vary in color from pale yellow to black. Whatever the color, all are mixtures of many compounds, the principal ones being—as in coal—hydrogen and carbon.

The crude oil is separated at the refinery into a number of mixtures of hydrocarbons. These are gasoline, kerosene, lubricating oils, fuel oils and asphalt with other products called "light ends," such as methane, ethylene, propane, and butane.

The items of most concern to us are fuel oils. The oil industry classifies oils used for heating as fuel oils numbered one through six. Number one oil flows easily and vaporizes readily. Number six flows with the most difficulty and is the hardest to vaporize. Number two oil, the type burned in residences and in small industrial boilers, vaporizes easily and requires inexpensive equipment to burn. It is high-priced per Btu, but the capital outlay is small.

The most popular fuel for large-scale industrial and marine use is number four or number six fuel oil.

Oil as a boiler fuel was first used on ships, where it proved to be superior to coal. The problems associated with coal, such as cleanliness of operation, disposal of ashes, spontaneous combustion in bunkers when at sea, and the labor required to fire it by hand made oil, by contrast, a natural fuel for the maritime field. It was common practice to load with coal in a home port and when in an oil-producing area such as the Middle East convert the boilers to oil fires and burn oil on the homeward trip. Other factors were a gain in space, since a ton of coal occupied 44 ft^3 compared with a ton of oil, which occupied 38 ft^3, and a greater heat value per pound of fuel: 18,500 Btu's for oil against 14,000 Btu's for coal.

Land-based boilers with ready access to coal were slow to change, but the economics of oil versus coal prevailed; large numbers of conversions were made and oil became the most widely used fuel.

As with all fuels, specific standards have been set to provide a guide to the user on the quality of the material burned. When burning fuel oil it is important to know the following:

(1) chemical composition

(2) flash point

(3) fire point

(4) heating value

(5) specific gravity

(6) viscosity

(7) pour point.

The chemical composition of fuel oil varies considerably depending on the grade of oil being burned. For the plant operator, the grades of fuel oil usually available are No. 2, No. 4, and No. 6 oils. Number 2 oil is a light fuel oil that can be burned satisfactorily without any particular pretreatment and is the most common fuel oil for domestic use and other small boiler needs. Number 4 fuel oil is heavier and must be preheated before burning. Number 6 oil, sometimes referred to as Bunker C, is the heaviest fuel oil and must be heated to be pumped and burned.

All of these fuel oils contain the following chemical elements:

Hydrogen	about 11 to 13%
Carbon	about 88 to 86%
Nitrogen } Sulfur	< 2%
Ash	< 1%
Water	Nil

As with all fuels, the heating value is obtained from the presence of hydrogen and carbon. The nitrogen, sulfur, ash, and moisture provide no heat value and reduce the efficiency of operation.

The most objectionable element in fuel oil is sulfur, which, when involved in a combustion process, forms sulfur dioxide and sulfur trioxide. Under Environmental Protection Agency regulations, the amount of sulfur permitted in fuel oils is strictly limited. Present rules call for 0.2% maximum in No. 2 fuel oil, 0.3% maximum in No. 4 oil and 0.4% maximum in No. 6 oil. This regulation, while reducing gaseous oxides of sulfur from escaping to the atmosphere, also has a side benefit in preventing excessive fireside corrosion in a boiler plant caused by the formation of sulfuric acid. Nitrogen in fuel oil adds nothing to the heating process but rather absorbs heat from the fuel being fired.

Ash in a fuel oil is the inorganic matter that remains after the liquid has been burned. Commercial specifications require this constituent to be listed only in No. 6 oil, all other oils just specify water and sediment. The amount of ash does not normally interfere with the burning of the fuel.

One constituent of most fuel oils that is of no value at all is water. Most specifications place water and sediment in the same category for an average combined total of two-tenths of 1%. For the astute operator it pays to sample each load of oil delivered for water content. When oil costs 40 cents a gallon, a 6000-gallon truck load of oil that is specified at 0.2% water contains 12 gallons of water. As all fuel oils are lighter than water, a year's delivery of fuel oil can leave a large residue of water in the fuel tanks of an average industrial plant.

The procedure for testing a fuel for water content involves mixing a sample of the fuel in benzol or gasoline and placing a portion in each of two counterbalanced centrifuge bottles with graduated bottoms. The two bottles are placed in opposite sides of a centrifuge and spun at 1500 r/min for 10 minutes. The water is flung to the base of the bottle, and the percentage of water content is shown by a graduated cylinder scale. Combined water and sediment are referred to in the industry as BS&W which means bottom sediment and water.

A method for determining the amount of water in a fuel storage tank is with a water-finding paste. The paste is applied to a stick long enough to reach the bottom of the tank. The stick is dropped down a tank opening and the presence and depth of water are determined by a change in color of the paste. Commercially available sampling cylinders can also be utilized to extract a sample of the BS&W from the tank. This device is a cylinder with a screw cap and a plunger-operated valve. When it is lowered into the tank, the plunger strikes the bottom of the tank, opening the valve, and allowing the contents at the bottom of the tank to enter the cylinder. When the sampling cylinder is raised, the valve closes, allowing the sample to be withdrawn.

The hydrogen in fuel oil averages 12% with a heating value of 2300 Btu's/lb. Carbon at 87% provides 16,800 Btu's/lb, for an average of 19,000 Btu's/lb.

The heating value of a fuel oil varies little for each grade, and a determination of the specific gravity and reference to the tables will indicate an approximate heating value. These values are shown in Table 3-1, As we have noted, the heating value of a fuel can be established by using a calorimeter to measure the heat released from the burning fuel. To calculate the net heating value of the fuel requires a knowledge of the exact percentage of hydrogen in the specific fuel. The calorimeters in use today are extremely accurate in their reporting of the calorific or heat value of a fuel. The calorimeter illustrated in Figs. 3-1 and 3-2 is called an adiabatic calorimeter because changes are effected within a thermodynamic system without gain or loss of heat. The calorific value obtained in a bomb calorimeter test represents the gross heat of combustion for that sample. This is the heat produced when the sample burns, plus the heat given up when the newly formed water vapor condenses and cools to the temperature of the bomb. In a boiler, this water vapor escapes as steam in the flue gases and the latent heat of vaporization which it contains is not available for useful work. It will therefore be seen that the net calorific value of a fuel obtained by subtracting the latent heat from the gross calorific value is an important factor in calculation of boiler efficiency. It is necessary to know the percentage of hydrogen in the fuel to calculate this latent heat. Table 3-1 shows that the heating value increases in Btu's/lb as the specific gravity decreases. The specific gravity of oil, as measured by a hydrometer, is referred to as the A.P.I. gravity. See Table 3-2.

In December 1921, the American Petroleum Institute, the U.S. Bureau of Standards and the U.S. Bureau of Mines adopted what is now known as the A.P.I. scale for official use in the American petroleum industry. This scale is based on the formula:

$$\text{A.P.I.} = \frac{141.5}{\text{S.G.}} \text{ at } 60\,°F - 131.5$$

To obtain specific gravity from a given A.P.I., the formula is:

$$\text{S.G.} = \frac{141.5}{\text{A.P.I.} + 131.5}$$

When hydrometers are graduated, this is done at a temperature of 60 °F. The readings at or about this temperature will therefore be

TABLE 3-1
Heating Value of Fuel Oils
National Bureau of Standards

Gravity			Heating value	
Degrees A.P.I. at 60°F	Specific at 60°/60°F	Density (lb/ per gal)	Btu/lb	Btu/gal
10	1.0000	8.337	18,540	154,600
11	0.9930	8.270	18,590	153,900
12	0.9861	9.221	18,640	153,300
13	0.9792	8.164	18,690	152,600
14	0.9725	8.108	18,740	152,000
15	0.9659	8.053	18,790	151,300
16	0.9593	7.998	18,840	150,700
17	0.9529	7.944	18,890	150,000
18	0.9465	7.891	18,930	149,400
19	0.9402	7.839	18,980	148,800
20	0.9340	7.787	19,020	148,100
21	0.9279	7.736	19,060	147,500
22	0.9218	7.686	19,110	146,800
23	0.9159	7.636	19,150	146,200
24	0.9100	7.587	19,190	145,600
25	0.9042	7.538	19,230	145,000
26	0.8984	7.490	19,270	144,300
27	0.8927	7.443	19,310	143,700
28	0.8871	7.396	19,350	143,100
29	0.8816	7.350	19,380	142,500
30	0.8762	7.305	19,420	141,800
31	0.8708	7.260	19,450	141,200
32	0.8654	7.215	19,490	140,600
33	0.8602	7.171	19,520	140,000
34	0.8550	7.128	19,560	139,400
35	0.8498	7.985	19,590	138,800
36	0.8448	7.043	19,620	138,200
37	0.8398	7.001	19,650	137,600
38	0.8348	6.960	19,680	137,000
39	0.8299	6.920	19,720	136,400
40	0.8251	6.879	19,750	135,800
41	0.8203	6.839	19,780	135,200
42	0.8155	6.799	19,810	134,700
43	0.8109	6.760	19,830	134,100
44	0.8063	6.722	19,860	133,500
45	0.8017	6.684	19,890	132,900
46	0.7972	6.646	19,920	132,400
47	0.7927	6.609	19,940	131,900
48	0.7883	6.572	19,970	131,200
49	0.7839	6.536	20,000	130,700

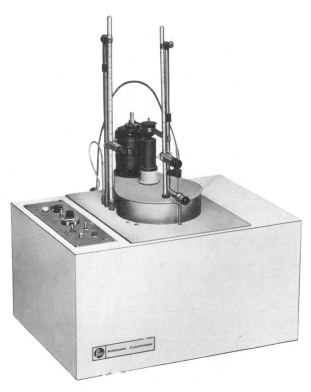

Figure 3-1. Adiabatic calorimeter for fuel analysis. *(Courtesy of Parr Instrument Company)*

accurate. If it is necessary to determine specific gravity at a temperature far removed from the temperature at which the instrument was graduated, it will be more accurate to correct the specific gravity at 60 °F to the required temperature by means of the density coefficient. The coefficient can be found in the volume correction tables for fuel oils.

Probably no other property of fuel oil is as important as viscosity. Number 2, 4, and 6 oils are burned by atomizing methods, and viscosity controls the effectiveness of atomization. Viscosity is a measure of a liquid's resistance to flow. The most common method for determining viscosity is to allow a sample of oil to flow through a capillary tube inserted into a small vessel that is surrounded by water at a constant temperature. This device is called a viscosimeter. The number of seconds for 100 cm³ of oil to flow through the tube shows the viscosity of the oil at the temperature of the test.

In the United States two different scales are in common use: the Saybolt Universal and the Saybolt Furol. The latter has a larger bore than the former and oil will flow through it in one tenth the time.

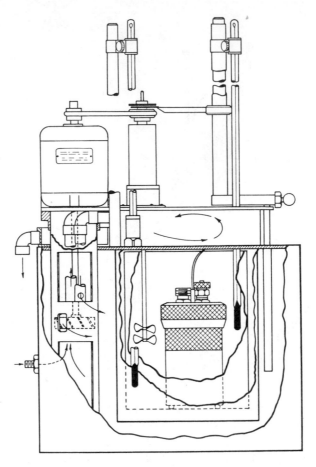

Figure 3-2. Cross-section of fuel calorimeter. *(Courtesy of Parr Instrument Company)*

TABLE 3-2

Heating Value of the Constituents of Hydrocarbon Fuels

		Btu's per lb
Carbon	C	14,000
Hydrogen	H_2	61,000
Sulphur	S	3,900

TABLE 3-3

Volume Correction Table for Petroleum Oils

	Group O*		
T†	M†	T	M
0	1.0211	155	0.9674
5	1.0194	160	0.9657

TABLE 3-3 (cont.)

		Group O*		
T†	M†		T	M
10	1.0176		165	0.9640
15	1.0158		170	0.9623
20	1.0141		175	0.9606
25	1.0123		180	0.9590
30	1.0106		185	0.9573
35	1.0088		190	0.9556
40	1.0070		195	0.9539
45	1.0052		200	0.9523
50	1.0035		205	0.9507
55	1.0017		210	0.9490
60	1.0000		215	0.9474
65	0.9982		220	0.9458
70	0.9965		225	0.9441
75	0.9948		230	0.9425
80	0.9931		235	0.9409
85	0.9914		240	0.9392
90	0.9896		245	0.9376
95	0.9879		250	0.9360
100	0.9862		255	0.9344
105	0.9844		260	0.9328
110	0.9827		265	0.9312
115	0.9809		270	0.9296
120	0.9792		275	0.9280
125	0.9775		280	0.9264
130	0.9758		285	0.9248
135	0.9741		290	0.9233
140	0.9724		295	0.9217
145	0.9707		300	0.9201
150	0.9691			

		Group 1*		
T	M		T	M
0	1.0242		30	1.0120
5	1.0222		35	1.0100
10	1.0202		40	1.0080
15	1.0181		45	1.0060
20	1.0161		50	1.0040
25	1.0140			
55	1.0020		155	0.9628
60	1.0000		160	0.9008
65	0.9980		165	0.9590

TABLE 3-3 (cont.)

Group 1*

T	M	T	M
70	0.9960	170	0.9570
75	0.9940	175	0.9551
80	0.9921	180	0.9532
85	0.9902	185	0.9513
90	0.9881	190	0.9494
95	0.9861	195	0.9476
100	0.9841	200	0.9457
105	0.9822	205	0.9438
110	0.9803	210	0.9419
115	0.9783	215	0.9401
120	0.9763	220	0.9382
125	0.9744	225	0.9363
130	0.9724	230	0.9345
135	0.9705	235	0.9326
140	0.9686	240	0.9307
145	0.9667	245	0.9289
150	0.9647		

Group 2*

T	M	T	M
0	1.0297	80	0.9901
5	1.0273	85	0.9876
10	1.0248	90	0.9851
15	1.0223	95	0.9826
20	1.0198	100	0.9802
25	1.0174		
30	1.0149	105	0.9776
35	1.0124	110	0.9752
40	1.0099	115	0.9727
45	1.0074	120	0.9702
50	1.0049	125	0.9677
55	1.0025	130	0.9652
60	1.0000	135	0.9627
65	0.9975	140	0.9603
70	0.9951	145	0.9578
75	0.9925		

*Group O applies to fuel oils of A.P.I. gravity of 60°F up to 14.9; Group 1 to gravity range from 15.0 to 34.9; Group 2 to gravity range from 35.0 to 50.9.

†T: observed temperature in degrees Fahrenheit

M: multiplier for reducing oil volumes to the basis of 60°F

Common practice is to express viscosities of No. 4 and No. 6 oils in Saybolt Universal at 100°F. Sometimes No. 6 oil viscosity is expressed in Saybolt Furol at 122°F. Furol viscosimeters are run at the higher temperature to reduce the time of testing. Other countries have different viscosity scales, such as the Redwood in Britain and the Engler in Europe. Conversion factors can be applied to adjust the readings.

Viscosity changes rapidly with temperature and is particularly important when handling a No. 6 fuel oil. Preheating renders the most viscous oil fluid enough to permit pumping and successful burning. A 300-SUS viscosity at the burner tip is desirable.

A relationship or balance exists between the viscosity and temperature of an oil for maximum combustion efficiency. If the temperature exceeds the optimum figure, the output of the burner will be decreased because of increased volume; if the temperature is below the ideal figure, the output could be diminished due to increased viscosity and poor atomization. For a given pressure the maximum output can be obtained only with a correct temperature/viscosity balance.

The temperature at which an oil burns correctly is called its fire point and obtaining this correct temperature can usually be determined only by trial and error during burning.

Determining the pour point of a fuel oil is a simple test used to express the lowest temperature at which the fuel oil may be exposed and still be capable of being pumped. The test consists of gradually lowering the temperature at which the oil momentarily holds a level surface, perpendicular to the test tube holding it, when the tube is tilted from a vertical to a horizontal position. Typical readings for a No. 4 fuel oil can be as low as −5°F.

The flash point of a fuel oil is the temperature at which the heated oil will give off a vapor that ignites when brought into contact with a flame. It is usual in fuel oil specification to see the flash point designated as P.M.C.C. followed by a certain temperature in Fahrenheit degrees. The designation P.M.C.C. means Pensky-Martin Closed Cup. In the closed flash test the sample of oil is heated in a covered vessel resting on a frame and air jacket casing. The dish is covered by a hood provided with suitable openings for a stirring rod, a thermometer, and a sliding shutter, through which a lighted taper is inserted at short intervals while heating the oil sample. When, on inserting the lighted taper, a pale blue flame is seen on the surface of the oil, the temperature is at the flash point.

Flash points can also be determined by an open cup method, whereby a quantity of oil is heated in an open vessel and a lighted taper passed over the surface as the oil is heated. When a pale blue flame or flash is observed on the oil's surface, the observed temperature is recorded as the ''open'' flash point.

If the oil in the open test continues to be heated, the oil will be found to flash, ignite, and continue to burn. The temperature at which this occurs is called the ''fire point'' of the oil.

It is important to be aware that the flash point of the oil in exposed fuel tanks can very easily be reached if close watch is not kept on storage tank temperatures. Fuel oil tanks that are steam or electrically heated and exposed to sunlight should be monitored closely to insure that temperatures remain below the temperature at which the volatile vapors are given off. If this occurs, the calorific or heating value of the oil can be reduced very rapidly, resulting in an apparent low boiler efficiency. Also be aware that high temperature in fuel storage tanks can cause a vapor lock in fuel-oil suction lines due to vaporizing of the fuel when exposed to the negative pressure created by the suction of the pump.

Every substance has a heat capacity peculiar to itself. This property of heat absorption is, as we know, called specific heat. The specific heat of fuel oil is between 0.4 and 0.5 Btu/lb. The application of a specific heat factor in fuel oil is of use when designing preheaters and is usually assumed to be 0.5 to provide a margin of safety when designing such equipment.

In the preceding discussion, we have covered the basic factors affecting the nature of fuel oil. It is a liquid that, when used as a fuel, we seek to convert to a gas because gas is theoretically the ideal boiler fuel. Compared with solid and liquid fuels, gas is extremely easy to handle. Its combustion rate may be readily adjusted and produces an even heat with no waste, soot, or ash.

GAS

There are three gaseous fuels generally available to the boiler operator: natural gas, coal gas, and propane.

The only naturally gaseous fuel is natural gas, which has an origin similar to that of petroleum and occurs wherever petroleum deposits are found. Some natural gas is available in the same well as crude petroleum, but most of it is produced from dry natural gas wells.

The composition of natural gas varies among wells, but the principal constituents are methane (CH_4) and ethane (C_2H_6). The Btu's per cubic foot also varies for the raw gas, but producers and gas companies mix the gas with CO_2 or N_2 to produce a standard heating value at 1000 Btu's/ft^3.

Coal gas is produced by the heating of bituminous coal in a closed container; air is excluded and the volatile elements are driven from the coal, leaving a solid residue called coke. Most coking processes provide for the recovery of coal tar, benzene, ammonia, and some light oils. The gas is piped to consumers for use in cooking stoves and heating. As with natural gas, the heating value is adjusted by the gas companies.

Propane (C_3H_8) and butane (C_4H_{10}), otherwise known as liquefied petroleum gas, are byproducts of the refinery processing of petroleum. These gases (and mixtures of the two) are compressed to a liquid and shipped in tank cars, tank trucks, and in steel cylinders. L.P.G. is utilized by many industrial concerns as a standby fuel when shortages of natural and manufactured gas occur. The gas has an exceptionally high heating value, 2558 Btu's/ft^3 for propane and 3210 Btu's for butane. To utilize these fuels in place of natural gas, a vaporizing unit consisting of a heater plus a gas/air mixing pump is required. The addition of the air to the high heat-value gas reduces the heating value to that of natural gas at 1000 Btu's/ft^3. Whether mixed with air or in their original state, propane and butane are heavier than air and, therefore, adequate ventilation is critical when these fuels are used.

For the standard boiler operation in the United States, all of the fuels just described constitute the bulk of normal use, and for this reason the combustion properties of these fuels are of concern to operating personnel. Much of the preventable waste in any power plant occurs in the combustion process; thus a precise knowledge of such processes is necessary to minimize energy losses.

Combustion is a chemical reaction, the constituents of the reaction being fuel and oxygen; as they interact, light is created and heat energy is liberated. Before combustion can occur, the temperature of the fuel must be raised to its ignition or kindling temperature. That is the temperature at which rapid oxidation occurs at such a rate as to generate heat faster than it can be dissipated. Although oxidation or burning occurs with the rusting of iron, the process takes place so slowly that no temperature rise is noticeable.

After part of the fuel has achieved its ignition temperature, combustion will proceed automatically. The heat released by the burning substance will then heat other parts of the substance until the entire mass is burning.

The principal constituents that unite with oxygen to produce combustion are: hydrogen, carbon, and sulfur. The other elements in a fuel do not add to the heating value and for all practical purposes may be ignored.

When a substance is burned, its combination with oxygen produces what is known as the "products of combustion." When two substances combine to form a product of combustion they join in definite proportions by weight, and a specific amount of heat energy is released. The amount of heat energy thus released is expressed in British thermal units, and the heating value of the fuel is so many Btu's per pound of fuel, except in the case of a gaseous fuel, which is expressed in Btu's per cubic foot.

In all boiler operations, the oxygen used in the combustion process is available from the atmosphere. The earth's atmosphere is a mixture of gases, in which oxygen and nitrogen are predominant. By weight, air contains 23% oxygen and 77% nitrogen. By volume, air contains 21% oxygen and 79% nitrogen. These percentages total 100% because we consider the other trace gases in the atmosphere as part of nitrogen and as negligible. It will be seen, therefore, that if 23 lb of oxygen is required for the combustion of a certain amount of fuels, 100 lb of air (oxygen and nitrogen) must be supplied to the furnace.

The complete combustion of carbon results in a product of combustion called carbon dioxide, with the liberation of heat. The carbon (C) and the oxygen (O_2) combine to form carbon dioxide (CO_2); the chemical formula is: $C + O_2 = CO_2$.

Amadeo Conte Di Quaregna Avogadro (1776–1856) an Italian physicist, advanced in 1811 the hypothesis, since known as Avogadro's law, that equal volumes of gases under identical conditions of pressure and temperature contain the same number of molecules. Since then, other physicists have determined that the number of molecules in the gram molecular volume is the same for all gases.

In general, all gaseous elements have two atoms in each molecule: the chemical symbol for oxygen is O_2, for nitrogen N_2.

According to Avogadro's law, therefore, if all the oxygen combined with all the pure carbon in the combustion process, the theoretical CO_2 content of a flue gas should be 21%.

Actually, the hypothetical maximum of 21% CO_2 is never attained in practice. Some unreacted oxygen always remains in the products of combustion. Oxygen can undergo chemical reaction with an element only when coming into actual contact with it. Practically, it is impossible to mix the air and fuel so thoroughly that every molecule of oxygen comes into contact with unburned fuel.

Another reason for not attaining perfect combustion is that all carbon may not burn to produce carbon dioxide. The intermediate stage of carbon combustion is carbon monoxide (CO), also a gas. The presence of carbon monoxide represents a serious loss of fuel. It should, therefore, be the aim of a boiler operator to keep carbon monoxide to an absolute minimum.

A molecule of oxygen contains two atoms, and a molecule of carbon monoxide contains one atom of oxygen. By Avogadro's law, the monoxide volume is double that of the oxygen from which it was formed.

As an example of the possible losses, suppose one pound of carbon undergoes incomplete combustion and instead of being burned to CO_2, it is only burned to CO. In Table 3-4 it will be seen that the burning of one pound of carbon to carbon monoxide releases 3960 Btu's but if burned to carbon dioxide, releases 14,150 Btu's. Therefore, for every pound of carbon which undergoes incomplete combustion, 10,190 Btu's will be wasted.

Thus far it has been assumed that the only fuel burned has been carbon. However, all commercial fuels contain hydrogen, and when it burns, hydrogen forms water and so consumes some of the oxygen supply. This further reduces the percentage of CO_2 in the flue gas. Since the water molecule contains one atom of oxygen, it too, causes an increase in gas volume by its formation. Water vapor condenses when the flue gases are cooled, so it can exist only at temperatures above the temperature of condensation.

The most widely used index of efficiency of boiler operation is the percentage of CO_2 in the flue gas. As noted, the theoretical percentage of 21 is not attainable in practice, but the nearer we approach that figure, the more efficient the operation.

The actual value of the carbon dioxide content of flue gas depends on the hydrogen in the fuel. The principal heat loss of combustion is the heat required to raise the temperature of the nitrogen and excess air. It is obvious that the lower the percentage of excess air, the less heat will be carried away up the stack and lost to the atmosphere.

TABLE 3-4
Heating Values

Carbon	→	CO	=	3,960 Btu's
Carbon	→	CO_2	=	14,150 Btu's
Hydrogen	→	H_2O	=	61,100 Btu's
Sulphur	→	SO_2	=	3,983 Btu's

The sensible heat that is thus lost may be calculated by the following formula:

$$H = 0.24 \, W \, (T_2 - T_1)$$

where H is heat in Btu's, W is the weight in pounds of the products of combustion for each pound of fuel burned, T_2 the temperature of stack gases, T_1 the temperature of the air supplied to the furnace, and 0.24 is the mean specific heat of the flue gas expressed as Btu's/lb.

The air and fuel are admitted to the boiler furnace at temperatures far lower than the heat of combustion. They leave the furnace through the stack at temperatures around 500 °F. The fuel supplied to a boiler is not all utilized to produce steam, and so the balance of the energy is wasted by the following means:

(1) unburned fuel
(2) combustibles in the flue gas
(3) heat carried away in the flue gas
(4) moisture in the fuel
(5) radiation and leakage.

To determine the actual losses involved in a boiler operation, it is possible to calculate all losses and arrive at a "heat balance" for the boiler installation.

The formula for boiler efficiency is:

$$\text{Efficiency} = \frac{\text{Btu's absorbed}}{\text{Btu's fired}}$$

This is true of all primary power sources, and a rough calculation of boiler efficiency can always be made. For example: A boiler is burning fuel oil at 19,000 Btu's per pound. If the heat utilized in producing steam is 12,500 Btu's, what is the boiler efficiency?

Solution:

$$E = \frac{12,500}{19,000} = 65\% \text{ efficiency}$$

The weight of air required to burn each pound of fuel can be calculated by the following formula:

$$W = 3.036\,C\,\frac{N}{CO_2 + CO}$$

W is the weight of air, in pounds, supplied to the furnace for each pound of fuel, C is the number of pounds of carbon in one pound of fuel and N, CO_2, and CO are the percentages by volume of the nitrogen, carbon dioxide and carbon monoxide determined by flue gas analysis.

In review, the most commonly used parameter for boiler efficiency is the percentage of CO_2 gas in the stack emissions. To determine this percentage an analysis of the flue gas can be made using a device for analyzing the gas.

By manipulating a water bottle, exactly 100 cubic centimeters of the flue gas is drawn into a graduated burette. By again manipulating the water bottle, this 100 cc sample is passed into and out of a container. This contains potassium hydroxide, an alkaline solution, which absorbs the CO_2 in the gas sample. After the CO_2 has been absorbed, the sample is passed back into the graduated burette and the level of the remaining sample read. The difference between the 100 original cubic centimeters and the level reading indicates the percentage of CO_2 in the stack gases. In a similar manner the gas sample is then passed into another container, containing potassium pyrogallate, which absorbs the oxygen in the gas. After removal of the oxygen, the gas is passed into a vessel, containing copper chloride, which absorbs carbon monoxide. The reduction in the volume of the gas after each absorption cycle indicates the percentage by volume of CO_2, O_2, and CO. The remainder of the sample can be considered as nitrogen (N_2).

The flue gas analysis does not reveal all the products of combustion—for example, water vapor is condensed and not measured. The Orsat apparatus measures the dry gas only. Therefore, it will be found that the nitrogen will always be greater than 79%, the percentage by volume of the nitrogen in the atmosphere.

Another device that is commercially available to check CO_2 and O_2 content of flue gas is called a Fyrite unit (Fig. 3-3).

The installation of CO_2 and O_2 recorders (Fig. 3-4) has become prevalent of late when the necessity to conserve fuel supplies has ·prompted considerable interest in the content of the flue gases, especially to reduce excess air. In general, close control of excess air, with a resultant high percentage of CO_2, indicates an efficient opera-

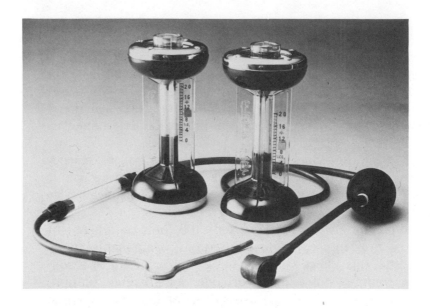

Figure 3-3. Fyrite O_2 and CO_2 stack gas analyses kit. *(Courtesy of Bacharach Instrument Co.)*

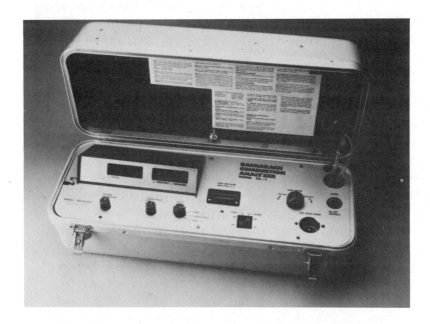

Figure 3-4. Bacharach digital combustion analyzer. *(Courtesy of Bacharach Instrument Co.)*

tion. (See combustion efficiency charts, Figs. 3-5, 3-6, and 3-7.) The results of a flue gas analysis appear in Fig. 3-8.

The output of a boiler is largely controllable by the operator, and the efficiency of the system depends on his skill and knowledge to maximize the heat used to evaporate the water in the boiler into steam.

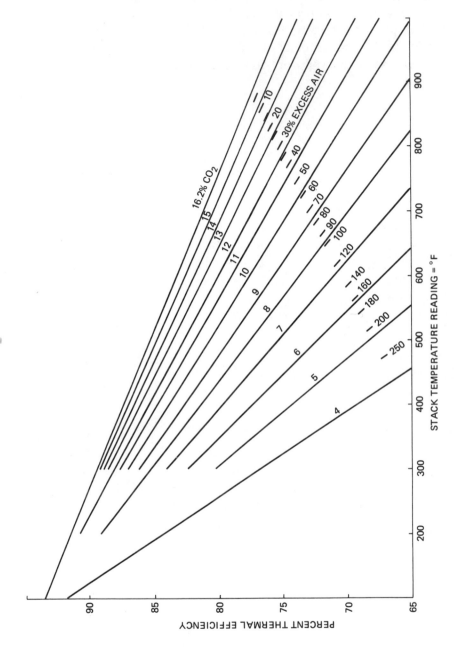

Figure 3-5. Combustion efficiency, No. 6 oil.

STACK TEMPERATURE READING = °F

PERCENT THERMAL EFFICIENCY

16.2% CO₂

30% EXCESS AIR

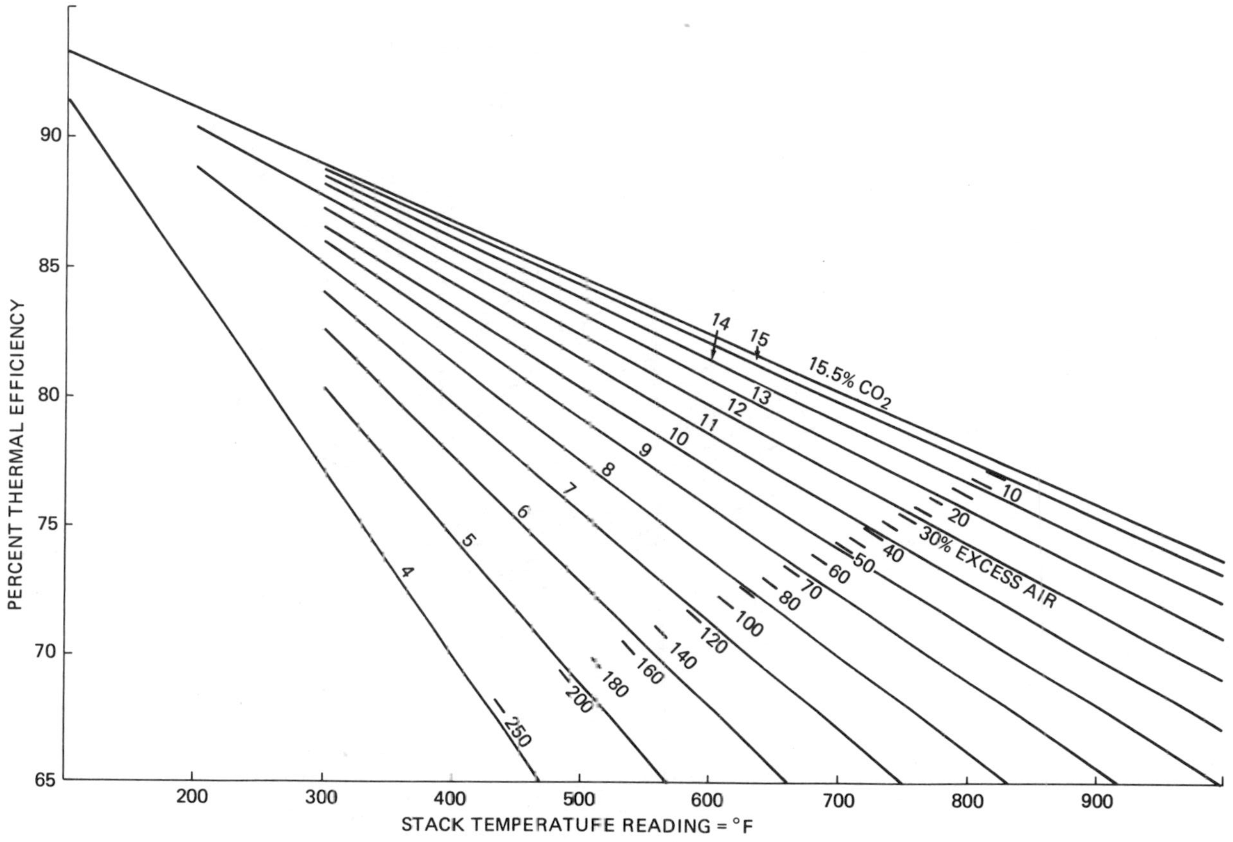

PERCENT THERMAL EFFICIENCY
90
85
80
75
70
65
200
300
400
500
600
700
800
900
STACK TEMPERATUFE READING = °F
14
15
15.5% CO_2
13
12
11
10
9
8
7
6
5
4
10
20
30% EXCESS AIR
40
50
60
70
80
100
120
140
160
180
200
250

Figure 3-6. Combustion efficiency, No. 2 oil.

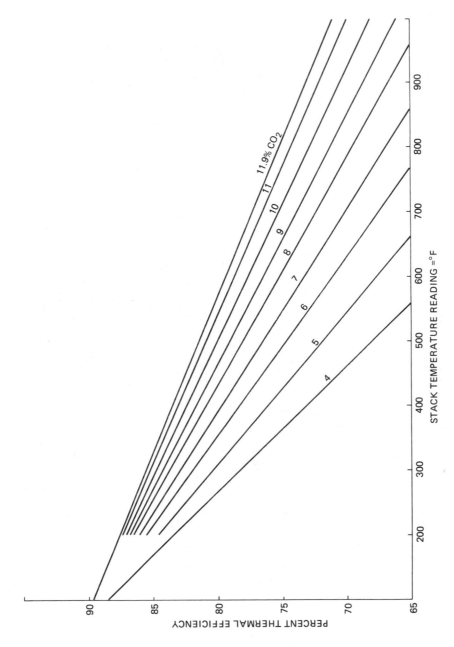

Figure 3-7. Combustion efficiency, natural gas (1050 Btu/ft^3).

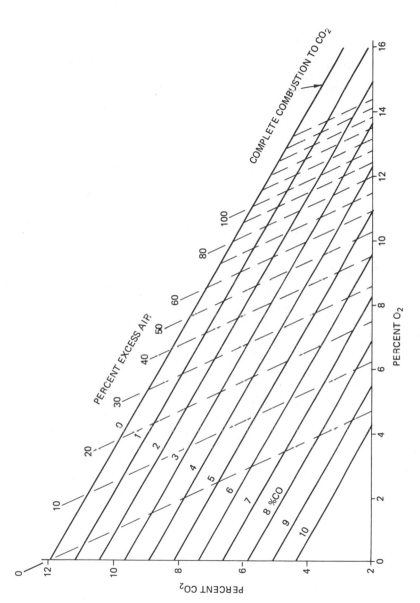

Figure 3-8. Flue gas analysis for natural gas (1050 Btu/ft³).

63

EXERCISES

1. Describe the burning characteristics and approximate heating values of anthracite and bituminous coal.

2. What is the purpose of a proximate coal analysis?

3. What determinations are made with an ultimate analysis?

4. Who was Pierre Dulong?

5. The ultimate analysis of a sample of coal has the following percentages:

Carbon	82.6
Hydrogen	2.45
Oxygen	1.9
Sulfur	0.4

   What is the heating value of this coal in Btu's per lb?

   Ans.:   13,364 Btu's per lb

6. Fuel oil is delivered to an industrial plant by tank truck. The oil specifications state that the oil contains 0.2% water. If the plant burns 25,000 gallons of oil per week, what amount, in gallons of water, is added to the fuel tanks annually? At 40¢ per gallon, what is the annual cost of the water?

   Ans.:   2600 gallons of water
   $1040.00

7. What is meant by the A.P.I. number of an oil?

8. What is viscosity? What effect does viscosity have on the pumping and burning of a fuel oil?

9. If you use a Pensky-Martin apparatus, what are you trying to determine?

10. Describe combustion. What is flash point, and fire point?

11. What is Avogadro's law?

12. Using Avogadro's law, what is the theoretical CO_2 content of a flue gas?

13. What gas should not be present in flue gas? Why?

14. Enumerate the heat losses that reduce boiler efficiency.

15. What is the efficiency of a boiler with a Btu input of 18,500 and an output of 13,000 Btu's?

    Ans.:   70.27%

16. If you were using a flue gas analyzer in the boiler room, what percentage of CO_2 would you want to see?

MULTIPLE CHOICE QUESTIONS

1. Anthracite is mostly:
   a. Hydrogen
   b. Carbon
   c. Calcium

2. The ultimate analysis of coal will tell you its:
   a. Ultimate value
   b. Chemical composition
   c. Heat content

3. The petroleum industry in the United States started in:
   a. 1909
   b. 1859
   c. 1890

4. When storing fuel oil it is important to know its:
   a. Heating value
   b. Flash point
   c. Specific gravity

5. A calorimeter that operates without heat loss is called:
   a. High efficiency
   b. Adiopose
   c. Adiabatic

6. Heating fuel oil will:
   a. Reduce its volume
   b. Reduce its viscosity
   c. Reduce its valence

7. The heating value of natural gas is:
   a. 100 Btu's/ft^3
   b. 1000 Btu's/ft^3
   c. 10,000 Btu's/ft^3

8. Air contains by weight:
   a. 23% O_2
   b. 21% O_2
   c. 22% O_2

9. To determine the composition of the products of combustion you can use:
   a. A fire test
   b. An Orsat test
   c. Good judgment

10. One pound of carbon burned to CO_2 will produce:
    a. 14,000 Btu's
    b. 14,500 Btu's
    c. 14,150 Btu's

4

Boilers

EVOLUTION

As we noted in Chapter 1, the boiling of water in an enclosed vessel to generate steam pressure has been practiced from ancient times.

When Thomas Savery and Thomas Newcomen developed the atmospheric engine and James Watt invented the condensing engine the pressures they used were extremely low. The early boilers were built with copper or wrought iron. Copper was a familiar material to the artisans of the late 18th century and was readily worked. Wrought iron was somewhat more difficult to handle and working it in thicknesses greater than ¼ in. was very difficult by early methods.

The evolution of the steam boiler followed the development of the steam engine. As the technology of the steam engine advanced, boilers were devised to provide the higher pressures needed.

Newcomen's boilers had been simple copper or wrought iron riveted spheres, set between brick supports with a grate underneath (Fig. 4-1).

The boiler devised by Watt was called a wagon boiler and consisted of a long cylindrical vessel set on walls with a fire grate beneath and flues on either side (see Fig. 4-2).

It did not take the engineers of the day long to realize that the longer the products of combustion were in contact with the shell of

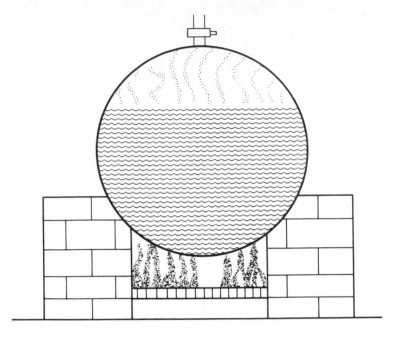

Figure 4-1. Early Newcomen boiler.

the boiler and the greater the surface area exposed to the heat, the more economical the boiler would be, producing more steam for less fuel.

The next advance was development of the internal fire grate, and this required an internal flue or fire tube. The "Cornish" boiler, invented in the United States, consisted of a long horizontal cylindrical shell, with a flue half the diameter of the outer shell running from end to end. At the front of this boiler the tube was enlarged to accommodate the grate and ash pit (Fig. 4-3).

To increase the heating surface the "Lancashire" boiler was devised, using two flues about one third the diameter of the outer shell, also with enlarged tubes at the front to accommodate the grate and ash pit (Fig. 4-4).

These boilers had flat ends and when pressure was raised the ends tended to buckle due to a combination of heat from the fire and the internal pressure of the boiler. To overcome the bulging of the flat ends, gussets were added to each end, but there is no record of the use of longitudinal staying. One form of stay that was used was called a Galloway tube (Fig 4-5), which helped support the internal flue. It also contributed to the circulation of the water in the boiler.

The Cornish and Lancashire boilers were used to accommodate the growing demand for steam and had typical plain-tube furnaces. The haystack boiler (Fig. 4-6) was developed to take advantage of the sphere and internal stays.

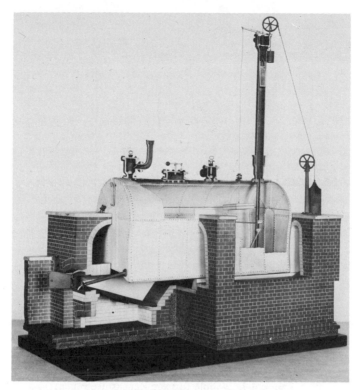

Figure 4-2. Wagon boiler. *(Courtesy of British Crown Copyright, The Science Museum, London)*

Figure 4-3. Cornish boiler, 1840. *(Courtesy of British Crown Copyright, The Science Museum, London)*

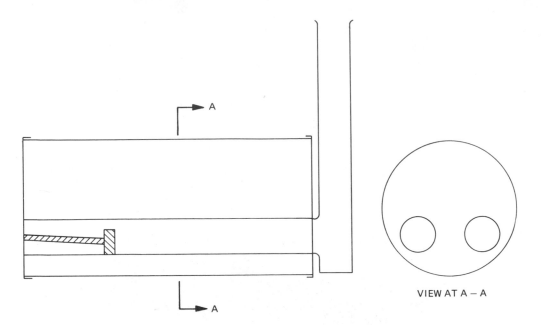

Figure 4-4. Double furnace Lancashire boiler.

Figure 4-5. Galloway boiler and setting. *(Courtesy of British Crown Copyright, The Science Museum, London)*

Figure 4-6. Haystack boiler, c. 1850. *(Courtesy of The Science Museum, London)*

Another problem was a way to absorb the expansion and contraction of the furnace tube. There were three basic ways in which this was accomplished, methods that are still used today. The first method divided the plain furnace into sections, each section being flanged at the ends, then riveted together with a soft iron caulking ring between the flanges (Fig. 4-7). The flanges stiffened the furnace and allowed for expansion, and the tightness of the joint was achieved by caulking the soft iron. This type of furnace is known as the Adamson ring furnace.

A further advance in furnace construction was the Bowling hoop furnace. This consists of plain sections riveted together with a convoluted section slightly larger in diameter (Fig. 4-8). This added strength to the furnace and also allowed for expansion.

The success of the Bowling hoop type of furnace with convoluted

SOFT IRON CAULKING RING

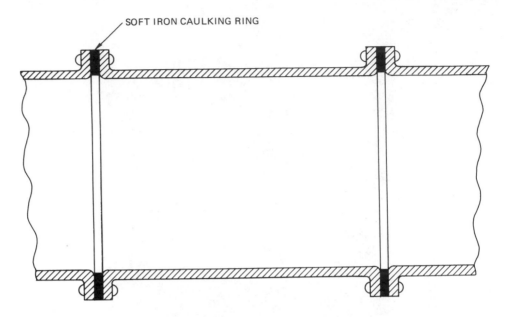

Figure 4-7. Adamson ring furnace.

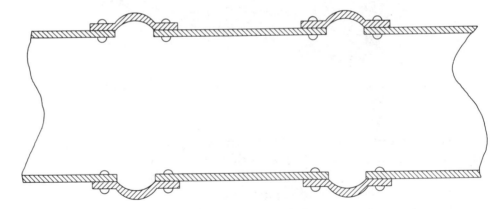

Figure 4-8. Bowling hoop furnace.

sections led to the corrugated furnace. In this furnace the plain tube was corrugated to provide for expansion of the metal (Fig. 4-9). The advantages of the corrugated furnace over the plain furnace are:

(1) greater strength with tne same dimensions
(2) better expansion allowance because of the corrugations
(3) increased surface area for the same length, achieving better evaporation.

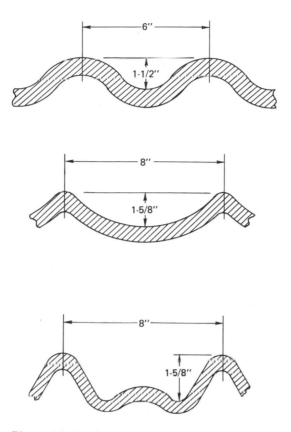

Figure 4-9. Section through corrugated furnaces.

The strength of a plain furnace depends on its length, diameter, and thickness squared ($L \times D \times T^2$). The strength of a corrugated furnace depends on its diameter multiplied by its thickness ($D \times T$).

As demands for steam increased, larger power plants, with bigger boilers, were required. The solution was to stand them on end, and thus a vertical boiler was built (Fig. 4-10).

The original vertical boiler, an adaptation of the Cornish and Lancashire boilers, became known as the vertical center flue boiler. Various adaptations of it were designed in order to increase the heating surface and obtain maximum value from the fuel burned.

The next innovation was to incline the fire tubes to allow more time for the products of combustion to stay in contact with the heating surfaces (Fig. 4-11). It was the essential prototype for the development of the locomotive boiler used by George Stephenson in his early railway work. It was the first example of what is known as a "dry-

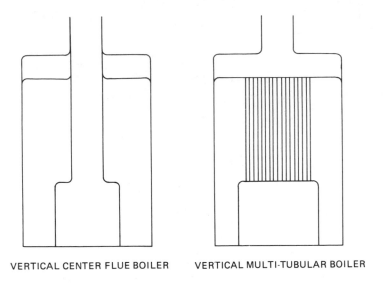

VERTICAL CENTER FLUE BOILER   VERTICAL MULTI-TUBULAR BOILER

Figure 4-10. Vertical boilers.

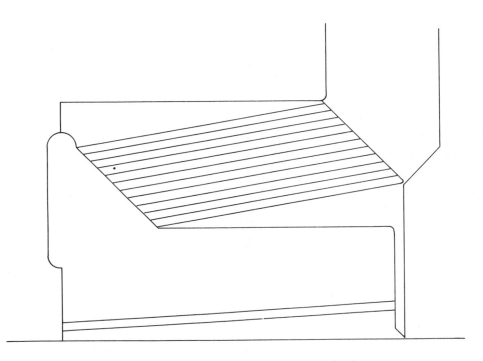

Figure 4-11. Horizontal boiler with diagonal tubes.

back'' boiler and was later developed into the pattern we know today as the horizontal return tube boiler or the H.R.T. (Fig. 4-12).

The next problem was to get more perfect combustion of the fuel to extract the full Btu value from the products of combustion. Normally the draft through the furnace was achieved by the rise of the hot gases in the chimney. Pictures of industrial plants in the 19th century show a profusion of tall chimneys with large volumes of black smoke pouring out over the surrounding countryside. The smoke, of course, was unburned fuel. With the addition of larger heating surfaces, the flue gases were cooler, reducing the draft through the furnace so the fuel would burn slower. Thus, it was possible to increase steam production only by increasing heat losses in the chimney.

George Stephenson had used a jet of exhaust steam on a locomotive to induce draft in a locomotive boiler as early as 1813, but to blow steam up a tall chimney was impractical, because the steam would merely condense and the moisture combining with sulfur in the fuel produced excessive corrosion. A jet of air from a fan injected at the base of the chimney was tried but it resulted only in a reverse flow, causing back pressure in the furnace and cooling of stack gases.

The eventual solution was to blow air from a fan into the ash pit through the fire bars of the grate and control the air flow with the opening and closing of the ash pit doors. This forced draft dramatically increased the steaming rate of boilers.

By the mid-19th century, boilers could be classified as vertical or horizontal and all were of the fire-tube type. This basic design of fire-tube boilers has been refined in many ways and is still in common use today.

As factories and railroads were utilizing steam power to replace human and windmill power, steam was replacing the sail in ocean transport. From the early 1700's to 1850, many plans were proposed to propel boats by steam. Some were brilliantly conceived, but not very successful. Robert Fulton is best known for his steam-driven vessel, the ''Claremont,'' which sailed in 1807 from New York to Albany in 32 hours. As better engines developed, marine boilers also improved.

The original boilers on ships were rather like the old dry-back boiler and pressures were very low, not above 25 pounds per square inch gauge. With the advent of the compound engine, steam pressures rose to 60 psi, and the Scotch marine boiler was designed. This type of boiler is still widely used today on land and at sea. For

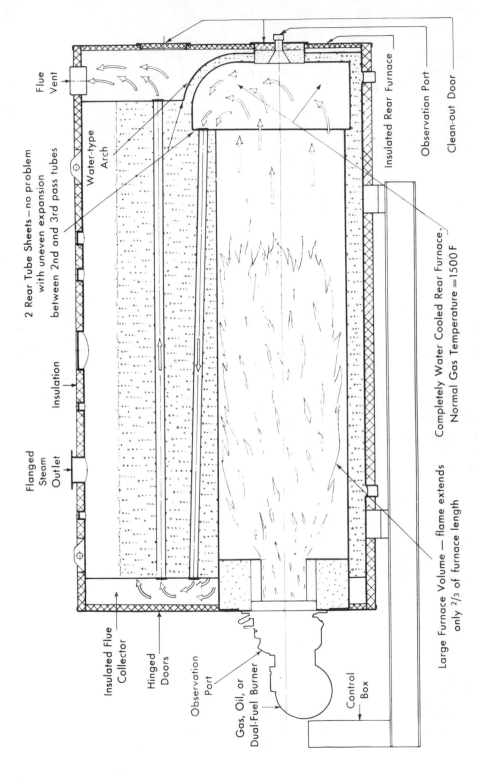

Figure 4-12. Atlas steam generator 3-pass boiler. *(Courtesy of North American Manufacturing Co.)*

Flue Vent

2 Rear Tube Sheets—no problem with uneven expansion between 2nd and 3rd pass tubes

Water-type Arch

Insulated Rear Furnace

Observation Port

Clean-out Door

Insulation

Flanged Steam Outlet

Completely Water Cooled Rear Furnace. Normal Gas Temperature = 1500 F

Large Furnace Volume—flame extends only 2/3 of furnace length

Insulated Flue Collector

Hinged Doors

Observation Port

Gas, Oil, or Dual-Fuel Burner

Control Box

use on ships, they have one to four furnaces and can be single or double ended.

The original conception of exposing a large amount of heating surface to the products of combustion culminated in 1867 with the invention of the Babcock water tube boiler.

As engines were developed, from compound to triple expansion to turbines, the necessity for higher pressures demanded a boiler that could generate pressures in excess of 300 psi. To achieve this in the horizontal return tube boiler and the Scotch marine boiler required excessively thick shells. The solution was the water tube boiler. The Babcock water tube boiler consisted of a bank of tubes inclined horizontally and connected at each end to headers. The ends of the tubes were staggered in the headers so that the hot gases flowed over the surface of each tube. The tube banks were baffled to allow the hot gases to contact all tubes in the bank. The tubes were 4 in. outside diameter (O.D.) in the lower banks and 1 $^{13}\!/_{16}$ in. (O.D.) in the upper banks. The front header terminated at the lower end in a valve where the boiler was blown out. The Babcock boiler (Fig. 4-13) was ahead of its time. Coal was cheap and the demand for steam was being met by already established boilers. With the invention of the electric light and turbogenerators, the necessity for fast steaming and economy of operation made the Babcock boiler critically useful.

In 1885 Allen Stirling designed a water tube boiler with three steam and water drums and two mud drums (Fig. 4-14). These were connected by tubes that were bent to approach the drums radially. The tubes were nearly vertical and tended to produce a high velocity of the mixture of steam and water bubbles in their rise to the upper drums. The steam and water drums were connected by short curved tubes, and the steam was extracted from the center drum. Stirling boilers are still manufactured and available with various combinations of drums, the most common being the tri-drum boiler. A number of water-tube boilers were under development in the 19th century, one of the most notable being the Yarrow boiler. Primarily designed for use on ships because of its lighter weight, and simplicity of design, it consisted of three drums: two water drums and a steam drum set in a triangular pattern. The original boiler had copper tubes but these were soon replaced by steel. After being proven at sea, many boilers of this type were adopted for land use.

Early in this century, boiler pressures in power stations were usually 200 to 300 psi. Design changes in equipment gradually required pressures up to 650 lb and up to 1000 to 1200 psi. These higher pressures

Figure 4-13. Babcock & Wilcox's marine-type water tube boiler, 1924. *(Courtesy of British Crown Copyright, The Science Museum, London)*

had to be met by modification in methods of construction. For pressures below 650 lb, riveted construction was used, but above that it became necessary to design forged seamless drums to withstand higher pressures. In 1925 electric arc welded steam drums were first introduced, but they were not generally used. In 1931 the American Society of Mechanical Engineers included welded joint standards in its Boiler and Pressure Vessel Code. Since that time, the welding of boiler drums has become accepted procedure with strictly enforced standards for the construction of any pressure vessel.

The construction of the modern boiler has made great advances over the original building techniques and methods used by the pioneers in boiler construction, when the joining of two plates was done with rivets. The early methods of construction were of the "cut and try"

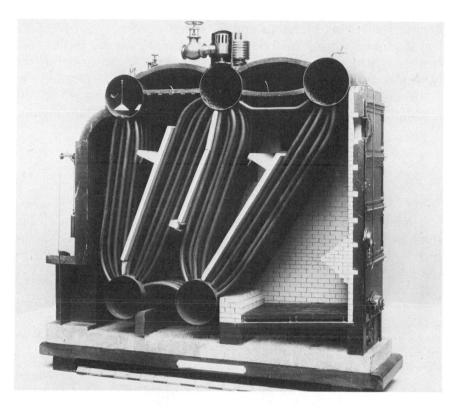

Figure 4-14. Stirling boiler. *(Courtesy of The Science Museum, London)*

school, but over the years these methods have been refined and standardized for all riveted and welded construction of boilers.

STRESS IN SHELLS AND JOINTS

When a boiler is steaming, or under load, the material of construction is stressed. There are two types of stress—tensile and compressive. The tensile strength of a boiler plate is its ability to withstand a pulling apart of the metal. Tensile strength is also known as the ultimate strength of the material. Shell plate for boiler construction averages 60,000 to 70,000 psi. If it is of a higher strength it becomes brittle and will not absorb expansion induced by heat.

During construction of a boiler, each plate is tested to conform to specifications. Specifications have been established by the American

Society for Testing Materials and the American Society of Mechanical Engineers for the methods of testing materials for boiler construction. Although most boilers are of welded construction, the principles of stress on shells apply to both riveted and welded shells.

In cylindrical boilers, the stress set up by the pressure on the longitudinal seam is equal to twice the stress on a circumferential seam.

$$\text{Stress} = \frac{\text{Forces causing break}}{\text{Forces resisting break}}$$

In a circumferential seam these forces are the area of the end of the boiler multiplied by the internal pressure in pounds per square inch and the circumference of the internal boiler shell multiplied by the thickness of the shell; thus:

$$\frac{\text{Area of head} \times \text{Pressure}}{\text{Circumference of head} \times \text{Thickness}} = \frac{0.7854D^2 \times P}{3.146 \times D \times T}$$

$$= \frac{D \times P}{4 \times T}$$

$$= \text{Circumferential stress}$$

Example

A pressure vessel is 36 in. in diameter with a shell thickness of 1 in. If the internal pressure is 500 psig, what is the stress on the circumferential seam?

$$\frac{D \times P}{4 \times T} = \frac{36 \times 500}{4 \times 1}$$

$$= \frac{18,000}{4} = 4500 \text{ lb/in}^2$$

On a logitudinal seam the same principles apply and the pressure tending to cause rupture on the longitudinal seam is equal to the length of the shell multiplied by the diameter of the boiler shell multiplied by the pressure.

The forces resisting the rupture are the two strips of metal equal to the length of the cylinder and the thickness of the metal. Thus:

$$\frac{\text{Diameter} \times \text{Length} \times \text{Pressure}}{2 \times \text{Length} \times \text{Thickness}} = \frac{\text{Diameter} \times \text{Pressure}}{2 \times \text{Thickness}}$$

$$= \frac{D \times P}{2\,T}$$

Using the same example as for the circumferential seam, the longitudinal stress would be:

$$\frac{D \times P}{2T} = \frac{36 \times 500}{2 \times 1}$$

$$\frac{18{,}000}{2} = 9{,}000 \text{ lb/in}^2$$

Therefore, for the same conditions of diameter, pressure, and shell thickness, the longitudinal stress will be twice the circumferential stress.

Note that the longitudinal pressure exerts a stress on the circumference and the radial pressure exerts a stress longitudinally. The strength of boiler shells therefore depends on the diameter and thickness and is independent of length.

In a test for tensile strength, a force of so many tons is applied to a strip of steel, forcing the steel to become elongated. If, when the force is removed, the strip of steel returns to its original dimensions, the stress put on the metal has been within the elastic limit of the steel. However, if the strip of steel remains elongated it is said to have exceeded its elastic limit and the steel has achieved a permanent set.

In testing a number of strips of steel, the safe stress can be assumed to be about half of this limit. We can then determine what is known as a *factor of safety.*

The factor of safety is determined as follows: Suppose, for example, 24,000 psi is found to be within the elastic limit of a steel plate and we assume that half of this is the safe working stress in a plate with a tensile strength of 56,000 psi.

Then

$$\frac{24{,}000}{2} = 12{,}000 \text{ lb}$$

which is the safe stress.

$$\text{Factor of safety} = \frac{56,000}{12,000} = 4.6$$

Before electric-arc welding was widely used in boiler construction, all boiler seams were riveted and caulked. As with shell and furnace plates, the rivets used had to meet certain standards, again set by the A.S.T.M. and the A.S.M.E.

In riveted boilers, internal parts were usually single-riveted (Fig. 4-15) and the circumferential seams and end plates were double-riveted (Fig. 4-16). Longitudinal seams were usually double butt strap riveted joints (Fig. 4-17).

The forces acting on a riveted joint tend to try and shear the rivets in the holes. The holes drilled or punched and reamed in the plate remove material from the plate edge and therefore weaken the plate.

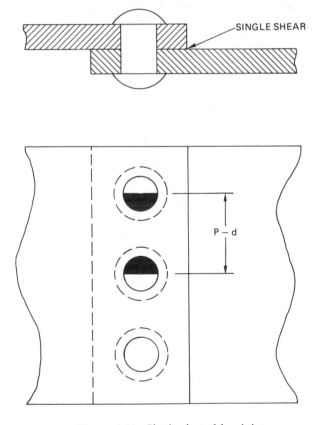

Figure 4-15. Single-riveted lap joint.

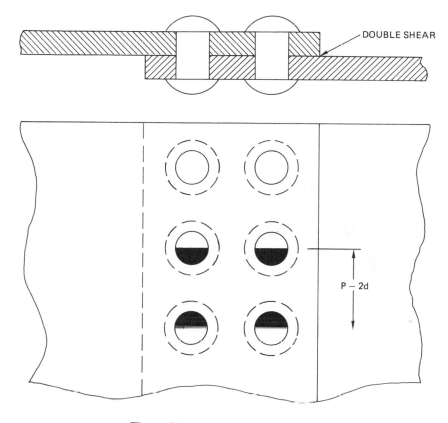

Figure 4-16. Double-riveted lap joint.

For this reason it is necessary to calculate the strength of the plate at the riveted seam.

Example

Assume a section of plate of *T* thickness has two rivet holes drilled in it, and designate the pitch between the center of the holes as *P* (Fig. 4-18). Let *d* be the diameter of the holes; comparing the solid plate with the plate with the holes drilled, the ratio would be:

$$P \times T: (P-d) \times T \text{ and } 100$$

that is, seam strength compared with solid plate.
Therefore:

$$\frac{(P - d) \times T \times 100}{P \times T} = \text{Seam strength}$$

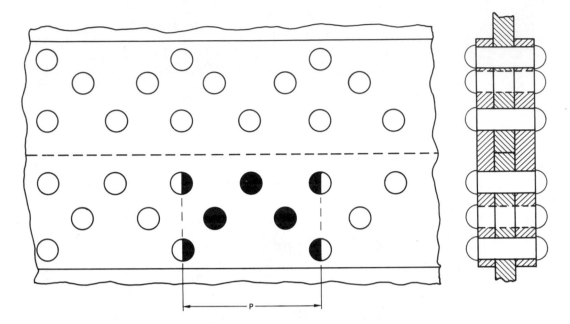

Figure 4-17. Double-riveted butt strap joint, 5 rivets to a pitch.

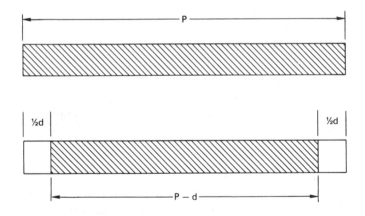

Figure 4-18. Rivet holes drilled in plate.

As T is common:

$$\frac{(P - d) \times 100}{P} = \text{Seam strength}$$

Rivets are fitted into holes and riveted tight, so that the rivet replaces the metal taken away from the plate by drilling the rivet holes. As in the example above to find the strength of a riveted seam:

$P \times T$: Area of rivets $\times$ Number of rivets

100: Riveted section compared to solid plate

Therefore:

$$\frac{d^2 \times 0.7854 \times \text{number of rivets} \times 100}{P \times T}$$

This formula assumes that both rivet and plate have the same strength but, in practice, the tensile strength of the plate exceeds the shear strength of the rivet.

The solid plate section for one pitch is equal to the pitch $\times$ the thickness of the plate and the rivet section area is equal to the rivet area multiplied by the number of rivets in one pitch. To apply the difference in tensile strength and shear strength between the plate and rivets the previous formula will now read:

$$\frac{d^2 \times 0.7854 \times \text{No. of rivets} \times 100 \times \text{Shear strength}}{P \times T \times \text{Tensile strength of plate}} = \text{Rivet section strength}$$

If double butt straps are used as in longitudinal seams the rivet section strength is doubled.

Single-riveted *lap joints* are used in furnaces and combustion chambers. Double lap joints are used in end plates and circumferential shell seams. Treble-riveted lap joints are used in center circumferential seams of long boilers. Double butt strap joints are only employed on longitudinal shell seams.

It is important for the boiler operator to be able to calculate the strength of various types of riveted joints found in boiler construction. All calculations are based upon restoring to a riveted joint the strength taken away from the solid plate by drilling of rivet holes.

Strength of a solid plate:

$$P \times T \times Ts$$

where P = Pitch or distance between the rivet holes

  t = Thickness of the plate

  Ts = Tensile strength of the plate

Using this formula, what is the percent strength of solid plate before drilling if we are going to construct a joint single-riveted lap having a rivet pitch of 4 in., using 1-in. rivets with plate ¾ in. thick?

Under the formula:

$$P \times t \times Ts = 4 \times 0.75 \times 55{,}000 = 165{,}000$$

The pitch will then be reduced by the diameter of 1 rivet; therefore:

$$(P - d) \times t \times Ts = (4 - 1) \times 0.75 \times 55{,}000$$
$$= 123{,}750 \text{ which is the strength of plate between rivets}$$

The shearing strength of 1 rivet in single shear, as there is only 1 rivet in the pitch, is equal to $s \times a$. Under this formula s is the shearing strength of the rivet and a is the cross-sectional area. Therefore:

$$s \times a = 44{,}000 \times 1^2 \times 0.7854$$
$$= 34{,}557.6$$

As a single-riveted lap joint is in single shear, the joint can fail by the crushing of the plate between the plate edge and the rivet hole. It is therefore necessary to calculate the strength of the plate between the plate edge and the rivet hole. In all single-riveted joints the distance from the plate edge to the edge of the hole must be equal to or greater than the diameter of the rivet. The crushing strength of the plate is determined by the formula:

$$d \times t \times c$$

where d = Diameter of the rivet or the distance between the edge of the rivet hole and the edge of the plate, whichever is greater

  t = Thickness of plate

  c = Crushing strength of mild steel, usually considered 90,000 psi

Therefore:

$$d \times t \times c = 1 \times 0.75 \times 90,000 = 67,500$$

For the particular joint cited:

Strength of solid plate $\quad\quad\quad\quad\quad$ = 165,000 psi
Strength of plate between rivet holes = 123,750.6 psi
Crushing strength of plate $\quad\quad\quad\quad$ = 67,500 psi

The lowest value is divided by the strength of the solid plate to give the joint efficiency. Therefore:

$$\frac{(s \times a):}{(P \times t \times Ts):} \quad \frac{34,557.6}{165,000} = 20.94\%$$

This of course is an extremely weak joint and would not be of practical use, but the method used to calculate riveted joint strength applies to all types of joints.

In double-riveted lap joints the joint may fail for the following reasons:

(1) Rivets may shear.
(2) The plate may crack between holes.
(3) The plate metal may be crushed between edges of plate and rivet hole.

Example

In a double-riveted lap joint the pitch is 3 $\frac{1}{16}$ in., diameter of the rivets is 1 in., thickness of the plate is $\frac{5}{8}$ in. and distance from the outer row of rivets to the edge of plate 1½ in. What is the joint strength? As before:

$$\text{Strength of solid plate} = P \times t \times Ts$$
$$= 3.0625 \times 0.625 \times 55,000$$
$$= 105,273.43$$
$$\text{Strength of plate between rivets} = (P - d) \times t \times Ts$$
$$= (3.0625 - 1) \times 0.625 \times 55,000$$
$$= 70,898.437$$

In this joint we have 2 rivets per pitch so the shearing strength is multiplied by 2. Thus:

$$2 \times s \times a = 2 \times 44{,}000 \times 1^2 \times 0.7854$$
$$= 69{,}115.2$$

and
$$d \times t \times c = 1 \text{ in.} \times 0.625 \times 95{,}000$$
$$= 59{,}375 \text{ psi}$$

$$\text{Joint \% strength} = \frac{59{,}375}{105{,}273.43} \times 100 = 56.4\%$$

It will be noted that as the plate thickness increases the strength of single- and double-riveted joints decreases.

In addition to riveting seams to construct the boiler, it is necessary to support the boiler components to prevent the internal pressure from distorting these parts. As in earlier days, it is necessary to support all flat surfaces in the boiler to prevent them from bulging when subjected to internal pressure. The cylindrical surfaces do not require stays as they do not change their shape under pressure.

In the horizontal return tube boiler the front and rear tube sheets are supported by through stays. These are long round bars threaded at each end and fitted with nuts and washers to secure the ends of both plates. In addition to stay bolts, a certain number of the smoke tubes are designated as stay tubes. These are tubes that have an increased wall thickness and are threaded at each end and fastened in the same manner as the through stay bolt.

Combustion chamber stays as used in Scotch marine boilers range in diameter from 1¼ in. to 1¾ in. These stays are screwed into the

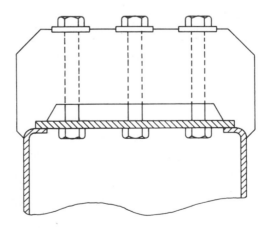

Figure 4-19. Combustion chamber girder stay.

plate to be supported and secured by nuts and washers. These stays are subjected to tensile stress that is sometimes accompanied by a bending stress. To identify fractured stay bolts a tell-tale hole is drilled by the center of the stay extending axially to a distance ½ in. beyond the inner face of the shell. Occasionally the nuts are omitted, the stays being screwed in and then riveted over.

The crown sheet on Scotch marine or other types of fire box boilers are supported by girder stays (Fig. 4-19).

The modern boiler, regardless of type, relies to a great extent on arc welding in its construction.

TYPES OF BOILERS

The welding on any pressure vessel must conform to A.S.M.E. rules and all welds are X-rayed to detect any flaws.

Modern design emphasizes greater heating surface and higher potential efficiency of operation. For example, the horizontal return tube boiler manufactured just a few years ago, with a 2-pass design, has given way to packaged designs incorporating 3- and 4-pass designs, which allows the hot gases more time to stay in contact with the heating surfaces and so transfer more heat to the water (Fig. 4-20).

Figure 4-20. 4-pass boiler. *(Courtesy of Cleaver Brooks)*

Heat transfer and total heating surface is the key to boiler efficiency and in modern practice 5 ft² of heating surface is considered to be the equivalent of one boiler horsepower. In coal-fired boilers, the grate area was used to estimate boiler horsepower and 10 ft² of grate area was equated with one boiler horsepower. In order to estimate boiler horsepower, by heating surface, all areas of the boiler exposed to heat from the fire and in contact with the boiler water should be taken into consideration.

Example

An H.R.T. boiler is 6 ft in diameter, 20 ft long, and has 70 tubes. The tubes are 2.75 in. outside diameter, 2½ in. inside diameter, and 18 ft long. What is the heating surface of the tubes?

Surface area of 1 tube $= \pi \times D \times$ Length

$$= 3.1416 \times 2.75 \times 18 \times 12$$

$$= 1866.11 \text{ in}^2$$

$$= \frac{1866.11}{144} \text{ft}^2$$

$$= 12.95 \text{ ft}^2 \text{ for 1 tube}$$

$$12.95 \times 70 = 906.5$$

Therefore, the heating surface of the tubes is 906.5 ft².

Calculations of the heating surface on any boiler require that all surfaces be taken into consideration.

The modern water tube packages boiler has the ability to develop more boiler horsepower than the older types. The water tube boiler was developed to take advantage of rapid water circulation and to produce steam at high pressure in a compact design (Fig. 4-21).

The Keeler-type CPM steam generator is a boiler of the water tube type (Fig. 4-22). The pressure unit consists of an upper water and steam drum, which extends the full length of the steel-cased insulated setting; a lower water drum; and two water wall headers, one on each side of the boiler, extending the full length of the setting, and connected at their rear ends to the lower water drum by means of short circulating tubes and to the upper drum by means of side water wall tubes. The feed water enters the boiler through an opening in the front head of the upper drum to which an internal feed pipe is attached. This pipe discharges the water at the rear end of the upper drum. From here the water flows down the rear bank of tubes

Figure 4-21. F.M. package boiler suitable for oil or gas firing. *(Courtesy of Babcock & Wilcox)*

into the rear end of the lower drum, then horizontally through the short circulating tubes into the water wall headers, forward and up through the side water wall tubes into the upper drum making a complete cycle of circulation. A second cycle of circulation occurrs in the main bank of boiler tubes, flowing downwardly from rear end of upper drum to rear end of lower drum, at the rear of the tube nest, and upwardly from the forward end of the lower drum into the upper drum directly above. This circulation occurs in a free and natural manner without the use of internal baffle plates, which are necessary in some other designs, and makes it possible to operate at high overloads with exceptionally dry steam and with no disturbance of the water level, which is normally at the center of the upper drum. The furnace water wall tubes do not depend upon the upper drum for their water supply but receive water directly from the lower drum. The steam generated in the furnace water wall tube is freely discharged directly into the upper drum without the use of intermediate headers.

The side walls of the furnace of the CPM boiler are water-cooled by tubes very closely spaced. The advantages of a water-cooled wall are several. By the installation of water tubes along the sides of the furnace, additional heat-absorbing surface is obtained without increasing the space occupied by the boiler. This heating surface is very effective, as it is exposed to the radiant heat of the fire, which is the best heat; and it will absorb more Btu's per square foot than heating surfaces in the main tubes of the boiler, which absorb heat by convection.

With a water-cooled furnace a higher percentage of CO_2 may be

(a)

(b)

Figure 4-22. (a) View of CPM generator with side casing tile and insulation removed; (b) a reproduction of an artist's drawing of a 500 hp CP steam generator fired by a single retort underfeed stoker. The design is patented in the United States (Patent No. 2,097,268) and in Canada (Patent No. 372,802). *(Courtesy of E. Keeler Co.)*

maintained than in a refractory-lined furnace. The result is a higher furnace temperature than can be safely carried in a refractory-lined furnace, yielding a higher boiler efficiency.

Another packaged type boiler is the Babcock & Wilcox "D" type construction. This design incorporates maximum furnace water walls including not only the side, roof, and floor but the rear wall as well. This boiler operates with a positive pressure in the furnace requiring a gas-tight casing.

For a high-capacity boiler of this type, efficient separation of the steam leaving the drum is essential to protect superheater tubes and keep turbine blades cleaner.

The Stirling boilers illustrated (Figs. 4-23 to 4-26) are the latest designs in a type of bent-tube boiler that was called the 4-drum boiler when originally introduced in the 1890's.

The 4-drum boiler had 3 upper drums and 1 lower drum. The lower drum was the only drum completely filled with water. Advances in design, to produce higher steam pressures and capacity, continue with all manufacturers.

CARE OF BOILERS

The proper care of boilers is of paramount concern to the operator. The following compact list briefly summarizes the major aspects of preventive maintenance.

WATERSIDE

Scale formation, pitting, corrosion, foaming, priming, wet steam, and fluctuating water level are the result of improper waterside care. Feedwater treatment, with proper blowdown procedures, is an important part of keeping the boiler heating surfaces free of scale.

Hot water systems are normally closed systems and should not require any makeup feedwater. However, some systems may have been installed in such a manner that system water is lost with regularity and makeup water is required. Feedwater treatment should then be used to prevent scale buildup and corrosion due to oxygen in the makeup water.

Oxygen Corrosion and Pitting

To protect your boiler from this type of deterioration, proper feedwater treatment is the only solution.

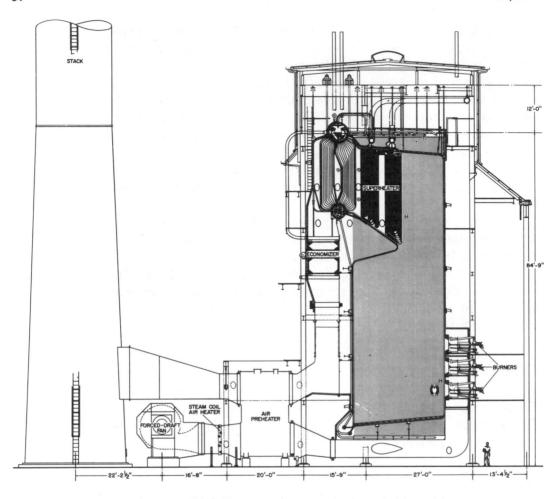

Figure 4-23. Oil fired power station steam generator (economizer and superheater). *(Courtesy of Babcock & Wilcox)*

Check the waterside surfaces (especially near the rear of the boiler) with lights and mirrors for any evidence of blisters, "pock marks," or erosion of metal surfaces.

If you note any of these conditions, your feedwater treating program needs immediate review and revision.

Scale Formation

Scale acts as an insulator and can result in overheating of the furnace, tubes, and tube sheets. This condition can cause tube leakage, tube end cracking, and other pressure vessel problems.

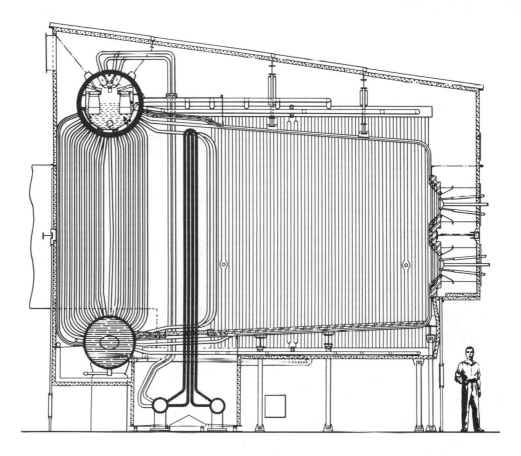

Figure 4-24. Power boiler to burn gaseous or liquid fuels. *(Courtesy of Babcock & Wilcox)*

Scale formation within any boiler is cause for immediate concern and action.

"Wet" Steam or Carryover

Wet steam or carryover can be caused by:

(a) high concentration of solids in the boiler due to lack of blowdown

(b) lack of proper feedwater treatment

(c) undersized steam line at boiler nozzle causing excessive steam velocities which "lift" water out of the boiler

(d) sudden or surge loads which hit the boiler due to quick-opening steam valves can cause instantaneous boiler overloading

(e) consistent boiler overloading due to increased plant load

(f) untrapped steam header.

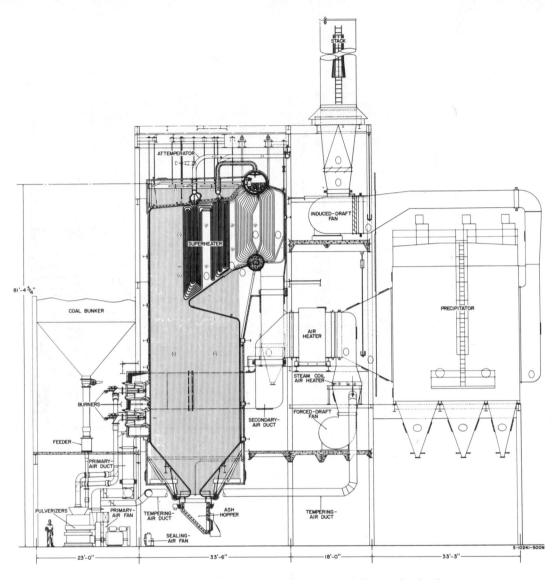

Figure 4-25. Stirling power boiler for burning pulverized coal. *(Courtesy of Babcock & Wilcox)*

Wetback and Firebox

Wetback boilers and firebox boilers have flat areas with flame or heat on one side and water on the other side. These flat areas have stay bolts or stay braces to reinforce the flat areas. Be sure these flat areas are free of scale and that the horizontal flat areas are free of an accumulation of mud or flakes of scale. Mud and scale buildup provide insulation, prevent heat transfer, and cause overheating.

Also, check stay bolts for corrosion or thinning.

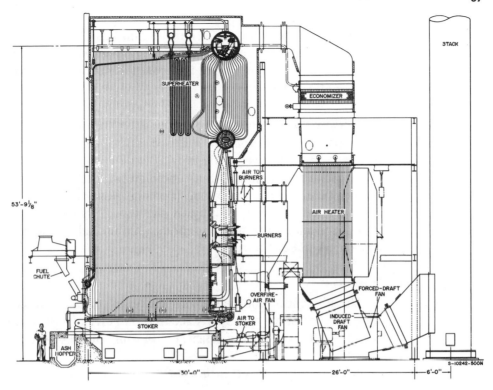

Figure 4-26. Stirling boiler. Stoker fired to burn different types of solid fuel. *(Courtesy of Babcock & Wilcox)*

Gaskets

It is important to maintain a tight seal at handhole covers. A slight leak will cause the boiler metal to erode, giving a poor seat for the gasket.

FIRESIDE

Furnace, Tubes, and Tube Sheets

Carefully check the fireside of the furnace and tubes for any evidence of blisters or "pock marks." This could indicate corrosion resulting from condensation of flue gases and formation of acid solution. This condition can be remedied by:

(a) maintaining a minimum boiler water temperature (control setting) of 170°F to avoid condensation of water vapor in the flue gas

(b) setting the controls so that the unit has the longest possible "on" cycle. Frequent cycling helps promote condensation

(c) reducing the fuel input if the unit is much too large for the actual load.

Tube Cleaning

Inspect the tubes for any evidence of soot deposits. Soot decreases heat transfer and lowers efficiency.

Install a thermometer in the boiler vent outlet. If the gas outlet temperature rises above normal, it means the tubes are dirty.

Evidence of heavy sooting in short periods of time could be a sign of too much fuel and not enough air and adjustment of the air–fuel ratio will be required.

Gaskets

Visually check door gaskets to make sure they are in good condition and that they are properly secured. An effective seal of fireside is required to prevent loss of efficiency, burning of gaskets, and deformation of the door steel.

Replace door gaskets if they don't give you a tight seal.

In summary, all of these maintenance practices will help to reduce fuel costs and insure that your boiler operates with high efficiency. The relationship between fuel-to-steam efficiency and fuel costs can best be illustrated as follows:

Example

Let's take a 200-horsepower boiler which burns No. 6 oil (costing 40¢ a gallon) yielding 150,000 Btu's per gallon. This 200-horsepower boiler produces 7000 lb of steam per hour, or 168,000 lb in a 24-hour day. In a 360-day year:

$$360 \times 168,000 = 60,480,000 \text{ lb of steam per year}$$

Assume your stack temperature is high and CO_2 is low. The efficiency of your boiler is 65%. You must burn approximately 600,000 gallons of oil at a cost of $240,000 to receive 60,000,000 lb of steam. However, if your boiler would operate at 80% efficiency, indicated by reduced stack temperature and high CO_2 in the flue gas, your costs would be much less. At 80% efficiency you would burn about 487,800 gallons of oil and your cost would drop to $195,120 to produce 60,000,000 lb of steam. Thus, the difference between 80% efficiency and 65% efficiency represents cost savings of more than $44,800 per year.

BOILER EFFICIENCY MEASUREMENT AND OPTIMIZATION

The primary objective of a boiler operation is to provide a continuous supply of steam at whatever pressure and temperature are suitable for the end use. The secondary objective is to provide such steam at the lowest possible cost. These objectives translate into "operational efficiency."

Throughout this volume stress is laid on the need for a high order of equipment maintenance to reduce operating costs, all part of operational efficiency. The need to burn fuel efficiently, obtaining all possible Btu values from the fuel burned, is a concern of the efficient operator.

The basic formula for boiler efficiency is:

$$\frac{\text{Btu's out}}{\text{Btu's in}} \times 100$$

This formula should be utilized on a daily basis for all operating boilers having the capability of measuring the amount of fuel used and the amount of steam produced.

Example

What is the efficiency of a boiler producing 20,000 lb/hr of dry saturated steam at 100 psig, burning No. 4 fuel oil with a heat value of 145,000 Btu's/gal. The feed water temperature is 213°F. The oil is burned at a rate of 120 gal/hr.

Referring to Table 2-1:

Heat required to convert 1 lb of water to steam $\quad$ = 708.5 Btu's

20,000 lb of steam would require 708.5 × 20,000 = 14,170,000 Btu's

120 gallons of oil $\qquad$ = 145,000 × 120 Btu's

$\qquad$ = 17,400,000 Btu's

$$\text{Efficiency} = \frac{\text{Btu's out}}{\text{Btu's in}} \times 100$$

$$= \frac{14,170,000}{17,400,000} = 81.4\%$$

Calculations such as these performed on a daily basis will provide a good general guide as to the efficiency of the boiler.

Improving Efficiency

The precise control of the air/fuel ratios and control of the chemical composition of the flue gases leaving the boiler are the key elements in any major improvement in boiler efficiency. The two conditions are dependent on each other. In single burner boilers a failure of the burner to produce the correct flame envelope for adequate mixing of the air and fuel can produce smoke or an excess of CO.

For many years the standard element used to measure efficiency was CO_2, but with the advent of the energy crisis a more definitive means was sought to evaluate the excess air, which contains 79% nitrogen that adds nothing to the combustion process; the measurement of oxygen content became a means of controlling the excess air in the stack gases.

Complete combustion of a fuel should react all hydrogen to water and all carbon to carbon dioxide with no free oxygen left over. In practice, however, there are always traces of oxygen (O_2) and carbon monoxide (CO). The measurement of the oxygen content to control excess air proved to be a not too accurate means of controlling excess air. Infiltration through idle burners, causing leaks, and other factors gave erroneous readings and did little to improve efficiency, especially on boilers with variable loads.

Experience over the last few years in numerous boiler operations has proved the value of monitoring CO in order to control excess air. With increased use of natural gas as a boiler fuel, the old method of adjusting the air by visual monitoring of stack emissions was no longer possible. Oil fuel and coal with insufficient air produce visible smoke in addition to colorless, odorless carbon monoxide. The development of portable and fixed CO analyzers has made use of the CO content of the stack gases a precise measurement of boiler efficiency.

To reduce excess air in a furnace, a systematic procedure must be followed. When a boiler is first put into service, a set of parameters is established for optimum performance at all load levels. These parameters should be available and referred to during any fine tuning of a combustion system. To reduce excess air in any furnace, the following procedure should be adopted:

(1) Set combustion controls on manual and bring boiler to a suitable firing rate for usable steam production.

(2) Adjust combustion controls, using CO analyzer for maximum efficiency at that firing rate. A reading between 100 and 300 ppm of CO indicates an oxygen content of 1% with excess air at less than 5%. These readings should be obtainable for gas or oil firing. With coal firing the oxygen and excess air numbers will double.

(3) Place combustion controls on automatic operation and continue to take readings at that level of firing. If the combustion controls are adjusted at the firing rate suitable for the steam being produced, there should be no change in the CO readings.

(4) The combustion controls should be placed on manual and a 10% reduction in firing rate established. Continue to adjust burner at progressive 10% reductions until low fire is reached. Continue to take readings of all established parameters at each step.

(5) After low fire readings are taken, bring firing rate to original point and proceed step by step to high fire.

(6) After these adjustments have been completed, vary the firing rate, throughout the entire range, making necessary adjustments for undesirable conditions.

When a high efficiency of operation has been obtained, constant follow-up, recording of operating conditions, and evaluation of results are mandatory for continued efficient operation.

EXERCISES

1. Why is the furnace of a boiler corrugated?

2. A boiler has a diameter of 60 in. and the plate thickness is ½ in. If the boiler has a working pressure of 150 lb/in., what is the circumferential stress in pounds per square inch?

Ans.: 4500 lb/in²

3. The longitudinal stress in a boiler drum is 10,000 lb/in². If the diameter is 40 in. and the plate is 1 in. thick, what is pressure in the boiler drum?

Ans.: 500 psig

4. What is the factor of safety? A boiler shell has a safe working stress of 25,000 lb/in². What is the factor of safety if the tensile strength of the plate is 56,000 lb/in²?

Ans.: 4.48

5. A piece of boiler plate is 6 in. long, 1 in. thick and has a tensile strength of 56,000 lb/in². Using the formula $P \times t \times Ts$, find the strength in pounds per square inch of this plate.

Ans.: 336,000 lb/in²

6. In Question 5, if ½ in. holes are drilled in the plate 4 in. center to center, what would the strength of the plate be between the centers of the holes?

Ans.: 168,000 lb/in²

7. A single-riveted lap joint is made of ¾ in. plate. The pitch of the rivets is 4 in. Rivet diameter is ¾ in. If shear strength of the rivet is 44,000 lb/in² and tensile strength of the plate is 56,000 lb/in², what is the efficiency of the joint?

Ans.: 11.5%

8. A double-riveted lap joint has a pitch of 3.5 in. The rivets are ⅞ in. diameter. The plate is ¾ in. thick. The distance from the plate edge to the outer row of rivets is 2 in. What is the % joint strength? Assume component strengths of 56,000, 44,000, and 90,000.

Ans.: 35.9%

9. What is the difference between a stay tube and a stay bolt?

10. A water tube boiler has a total heating surface of 3000 ft². What is its nominal boiler horsepower?

Ans.· 600 bhp

11. What advantages does a water tube boiler have over an H.R.T. boiler?

12. A boiler produces 5000 lb/hr of dry saturated steam. Boiler pressure 80 psia. Feed water 228°F. Fuel is 140,000 Btu's/gal. Boiler efficiency is 86%. How many gallons of oil will be burned per hour?

Ans.: 20.3 gal

13. What is enthalpy?

14. Describe in your own words the result of low excess air.

MULTIPLE CHOICE QUESTIONS

1. To support the flues and aid circulation, early boiler designers used:
   a. Galvanic tubes
   b. Galloway tubes
   c. Galling tubes

2. The strength of a corrugated furnace depends on its diameter and:
   a. Thickness
   b. Length
   c. Corrugations

3. The Stirling boiler was designed in:
   a. 1785
   b. 1885
   c. 1905

4. Boiler shell plate of too high a strength will become _____ under load:
   a. brittle
   b. tensile
   c. malleable

5. The stress in a pressure vessel varies _____ as the diameter:
   a. directly
   b. inversely
   c. conversely

6. Single riveted lap joints are used for:
   a. All boiler construction
   b. In furnaces and combustion chambers
   c. When shearing is not a factor

7. Flat surfaces in boiler construction:
   a. Are not allowed
   b. Must be stayed
   c. Do not happen often

8. Telltale holes are drilled into:
   a. Rivets
   b. Combustion chambers
   c. Stays

9. An H.R.T. boiler is also called:
    a. A high return tube
    b. A fire tube
    c. A water tube

10. An economizer is added to a boiler to:
    a. Reduce pollution
    b. Increase efficiency
    c. Increase steam temperatures

5

Boiler Accessories

SAFETY VALVES

The boiler with its drums, headers, tubes, grates, fireboxes, and walls is useless without adding all the accessories that make it function in the desired manner. These accessories make a boiler functional.

The list of boiler accessories includes—but is not limited to—valves, cocks, plugs, gauges, burners, pipes, dampers and stacks.

The most important valve on the boiler is the safety valve. The primary purpose of the safety valve is to limit the internal pressure of the boiler to a point below its safe working pressure.

The original safety valve in regular use was known as the "dead-weight" safety valve and consisted of a series of weights set on top of the valve, the weight depending upon the pressure to be retained (see Fig. 5-1). An improvement on this valve was the introduction of a lever-type dead-weight safety valve, which combined a lever, fulcrum, and weight to provide a variable pressure relief valve (see Fig. 5-2).

Marine applications made necessary use of a spring-loaded safety valve. The rolling of a vessel caused the center of gravity of the weights to vary, as well as the pressure at which the valve would lift (see Fig. 5-3).

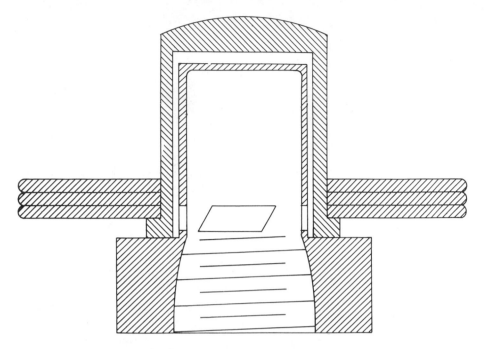

Figure 5-1. Cross-section of adjustable dead-weight pressure relief valve.

Figure 5-2. Dead-weight safety valve used on Cornish boiler, c. 1850. *(Courtesy of British Crown Copyright, The Science Museum, London)*

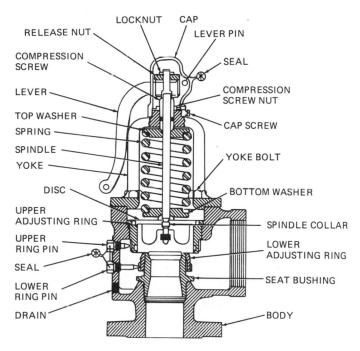

Figure 5 3. Typical safety valve construction details

To determine the load on any safety valve, multiply the valve area by the pressure in the boiler. To determine the pressure, divide the load on the valve by the valve area.

Example

A boiler has to operate at 60 psi. The diameter of the safety valve is 6 in. What load is needed on the valve?

$$6^2 \times 0.7854 \times 60 = 1,696.46 \text{ lb}$$

Example

The load on a dead-weight valve is 900 lb. The valve diameter is 2 in. What is the pressure in square inches that the valve will lift at?

$$\frac{900}{2^2 \times 0.7854} = 286.47 \text{ lb/in}^2$$

To calculate the load on the lever-type safety valve, the length (l) of the lever must be taken into account.
Thus:

$$l \times w = A \times \text{Boiler pressure}$$

NOTE: A is the area of the valve. Area $\times$ Boiler pressure = Load.

Therefore:

$$\frac{L \times w}{1} = \text{Load and} \quad \frac{\text{Load}}{A} = \text{Pressure in boiler}$$

Thus:

$$\frac{1 \times \text{Load}}{L} = W \quad \text{and} \quad \frac{1 \times \text{Load}}{W} = L$$

Example

$$L \times W = l \times \text{Load}$$

$$14 \times 100 = 6 \times \text{Load}$$

$$\frac{14 \times 100}{6} = \text{Load}$$

$$466.6 = \text{Load}$$

$$\text{Boiler pressure} = \frac{\text{Load}}{\text{Area}} = \frac{466.6}{8} = 58.33 \text{ psi}$$

In this example the weight of the lever and of the valve and spindle have not been considered. If these factors are to be allowed for, we have to know:

(1) the center of gravity of the lever
(2) weight of the lever
(3) weight of the valve and spindle.

In the previous example assume that the lever weighs 7 lb and its center of gravity is 12 in. from the fulcrum. The valve and the spindle weighs 5 lb. What is the load on the valve and the pressure per square inch?

$$\frac{12 \times 7 \text{ lb}}{6} = 14 \text{ lb more to the load}$$

$$66.6 \times 14 = 480.6 \text{ lb}$$

The weight of the valve and spindle is direct weight and is added directly.

$$480.6 + 5 = 485.6 \text{ total load in pounds}$$

$$\text{Pressure per square inch} = \frac{485.6}{8} = 60.7 \text{ psi}$$

These calculations illustrate the importance of considering all factors when determining the relieving of pressure of a dead-weight safety valve to prevent loading a boiler beyond its maximum allowable working pressure (MAWP).

Most commonly today, boilers are fitted with the spring-loaded pop-type safety valves. The A.S.M.E. code stipulates that the safety valve capacity of every boiler shall be such that the safety valve or valves will relieve all the steam that can be generated by the boiler without allowing the pressure to rise more than 6% above the MAWP or more than 6% above the highest pressure at which any valve is set. It is also required that one or more safety valves be set at, or below, the maximum allowable working pressure.

To insure that the foregoing conditions prevail, an "accumulation test" is conducted by a boiler inspector. It is a common practice for the inspector to bring test gauges to the site and connect them to the boiler steam pressure gauge piping, in order to secure an accurate gauge pressure reading. All boiler steam outlets are closed off and the boiler is operated at high fire. The safety valves lift and the boiler is operated in this fashion for 20 minutes. The gauge pressure is watched during this period to ensure a rise in pressure no greater than 6%.

All safety valves must be attached to the boiler steam space in an approved manner with no means of isolating such valves from the steam space. The discharge pipe must also be open-ended to the atmosphere with no means of closing the discharge pipe.

Safety valves must be fitted with a lifting device, which must be arranged so that the lifting device can be worked by hand from an accessible place free from the danger of blowing steam.

It is important that the valve body of a safety valve have a drain at the lowest point on the body above the valve seat to allow any water to drain to the atmosphere.

The action of all safety valves is to open at the set pressure and close when the pressure has been reduced to a few pounds below the set pressure. The difference between the set pressure and the closing pressure is called the "blowdown" of the valve. This is accomplished by using extended wings on the valve so that when the valve lifts off, it sets a greater area that is exposed to the steam pressure causing the valve to stay open beyond the set pressure.

Example

A safety valve is set to lift at 200 psi. The spring load is 1000 lb. What is the valve area?

$$\text{Area} = \frac{\text{Load}}{\text{Pressure}} = \frac{1000}{200} = 5 \text{ in}^2$$

When the valve is open and blowing the exposed area increases to 5.25 in², at what pressure will the valve close?

$$\text{Pressure} = \frac{\text{Load}}{\text{Area}}$$

$$\text{Pressure} = \frac{1000}{5.25} = 190.47 \text{ psi}$$

This blowdown is adjustable by setting the adjusting ring to cover or expose the ports in the seat.

The number of safety valves required on power boilers is determined by the gauge pressure and the boiler heating surface in square feet and is based on the formula:

$$A = \frac{HV}{420}$$

where A is total area of opening in square inches

H is boiler heating surface

V is specific volume of steam in cubic feet per pound at MAWP.

Example

A fire tube boiler has 1000 sq. ft. of heating surface and a MAWP of 125 psig. The specific volume of steam (from steam tables) is 3.22.

$$A = \frac{HV}{420} = \frac{1000 \times 3.22}{420}$$

$$= 7.667 \text{ in}^2$$

This means that the safety valves and the connections in the boiler must each have an open area of 7.667 square inches.

Good maintenance of the safety valve or valves on any boiler is essential for safe operation. The practice of easing the safety valves on their seats with the lifting lever at regular intervals is necessary to ensure that the valve will lift when required.

When inspecting a boiler, check the outlets to the safety valves to make sure that no deposits have accumulated in the outlet to either block the opening or freeze the valve to its seat. Priming a foaming boiler can plug safety valve outlets. If a safety valve is leaking it should be removed from the boiler and thoroughly overhauled by in-plant personnel or an outside concern. If the personnel or facilities are not available in the plant for proper overhaul, an outside vendor should be engaged; it is incumbent upon the plant personnel to be present when the outside vendor tests the valves to make sure the safety valves will lift at their set pressure.

To dismantle a safety valve for overhaul or repair, first remove the sealed cap and disengage the easing lever. Remove all tension on the spring by unscrewing the adjusting nut. The lock nut and adjusting nut should be center-punched to mark their positions and the number of turns to free the adjusting nut should be counted so that the valve can be reassembled in its original position. Remove valve cover, spring spindle, and valve.

Examine the spring for any signs of corrosion or cracks, and clean thoroughly. Examine the spindle for any corrosion or cracks. The spindle should be suspended from a line and tapped lightly to produce a clear ring. If the spindle does not ring clear, suspect a fracture in the spindle. Examine the valve and seat, to make sure blowdown parts are clear. It is not necessary to move the adjusting ring unless a change in the blowdown is to be made. Using a fine grinding paste, apply fine lapping paste in the seat and valve and check both for any marks. If all the parts are clear, reassemble in a reverse procedure from disassembly. In most jurisdictions it is necessary to notify either state or insurance company inspectors that you have had the valve open and they will check the setting and accumulation of the valve before resealing.

THE WATER COLUMN

While the safety valve has been called the "most important valve" on the boiler because of its ability to prevent excessive pressure, the water column, with its gauge glass, further ensures safe operation by allowing the operator to see how much water is in the boiler. Every manual on boiler operation, under all circumstances, gives the following instruction: The first duty of a boiler operator on taking over the operation is to check the level of water in the gauge glass.

The water column is a simple device, but a thorough understanding of its function and the means to ensure its accuracy is essential in boiler room operation.

Refer to Figs. 5-4 and 5-5: To ensure that the pipe to the steam side of the boiler is clear, close valves *C* and *D* and open drain *E*. If

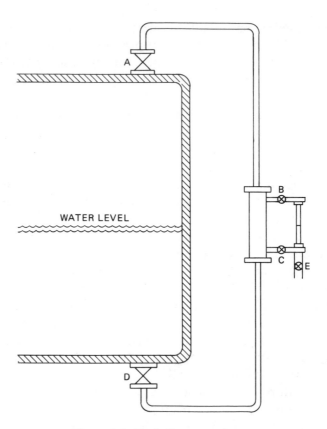

Figure 5-4. Typical gauge column.

steam blows, the passage is clear. To ensure that the pipe to the water side of the boiler is open, close valves *A* and *B* and open drain *E*. If water blows, the passage is clear.

If, after the gauge glass is blown, water fails to return to the glass, either valve *D* or *C* is blocked. To test for *D*, close valves *A* and *C* and blow through *D* and *B* to drain *E*; if water blows through, *D* is clear. To test *C*, shut valves *B* and *D* and blow through *A* and *C* to drain *E*; if steam blows *C* is clear.

If after blowing the glass the gauge glass fills completely, the steam side valves are plugged. To locate which valve is plugged, shut *D* and *B* and blow through *A* and *C* to drain *E*. If steam blows out, *A* is clear. To test *B*, shut *A* and *C*, blow through *D* and *B* and drain *E*. If water blows, *B* is clear.

If a boiler has been left unattended and the glass has not been blown for some time, the level in the gauge glass will be lower than in

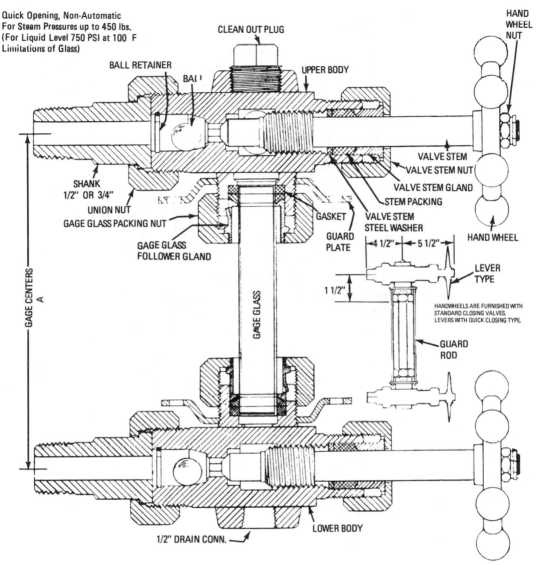

Figure 5-5. Cross-sectional view of water column cocks. *(Courtesy of Eugene Ernst Products)*

the boiler. Its volume has been reduced because the water in the glass is cooler than that in the boiler.

Restriction in either steam or water connections on the boiler to the connections on the water column will tend to show an erroneous reading on the glass. If the steam side is restricted or throttled in any way, the level in the glass will show a higher reading than the water that is actually present. Restriction or throttling in the water side will also cause a higher reading.

After blowing the glass and observing that the water has returned to the normal operating level, open and close the drain valve quickly. This should cause the level in the glass to jump. What you see is called a "lively" glass and should be the normal operating condition.

Incorporated in the water column are try-cocks. They are for use when the gauge glass is broken and there is no other way to determine where the boiler water level is located. (See Fig. 5-6.) Opening the try-cocks, the bottom one first, ascertains where the water level is. If water is located at the bottom cock, open the middle cock to check further. It must be emphasized that failure to locate the water level necessitates a complete shutdown of the boiler immediately. In practice, if the boiler gauge glass shows empty, shut off the fire immediately. If the gauge glass shows full, it is also correct practice to shut off the fire until the correct level can be ascertained. "Priming the boiler" caused by too high a water level, resulting in water leaving the boiler via the steam lines, can do considerable damage by causing water hammer in the headers and mains.

If it is not possible to open the try-cocks, an old alternative is to soak a rag in water and wipe it down the water column. The part

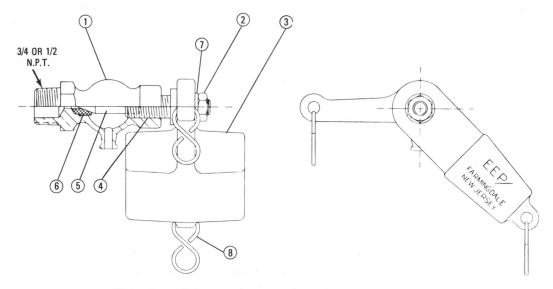

Figure 5-6. Cross-sectional view of a try-cock. *(Courtesy of Eugene Ernst Products)*

of the column in steam will dry very quickly; the section with water in it dries relatively slowly.

Also incorporated in the water column are various devices to warn of high and low water. An internal float opens a valve on high or low water to admit steam and produces a whistle that will give the operator audible warning of the condition before the low water level shuts down the fire or the high water causes the boiler to prime.

To avoid a gauge glass failure, it is good practice to replace the glass every 3 to 6 months. The action of the steam on the glass erodes the glass at the steam end, and eventually weakens it to the point where it will break.

For high-pressure operation (in excess of 250 psi) a flat gauge glass is used (Fig. 5-7). The columns are made with forged carbon and alloy steels or stainless steel. The glass is flat with a prismatic face and shows the steam as silver, the water as black. These glasses are protected with a mica shield to prevent erosion of the glass and are much longer lasting and more durable than the cylindrical glasses.

To install the gauge, follow the directions for tightening procedures carefully, since failure to do so will result in a cracked glass.

The water column is the operator's eyes inside the boiler; without a reliable gauge column, the operator is blind. The importance of close attention to the gauge cannot be overstressed.

GLOBES, GATES, AND CHECKS

The types of valves used, whether they are attached to the boiler or employed as line and header valves in a power plant operation, are of three general types: globe valves, gate valves, and check valves. The type of valve used will depend upon the type of service required since each valve has a specific function.

It is important to select the correct valve for each use on the basis of knowledge of necessary pressure, rate of flow, and operating temperature.

The globe valve is among the most versatile of valves and can be used to regulate or throttle flow or stop the flow altogether (see Figs. 5-8 and 5-9). Using a globe valve will result in a high pressure drop so the globe should not be used where pressure drop is a consideration. The most important use of the globe valve in the boiler room is as the main stop valve on the steam outlet from the boiler.

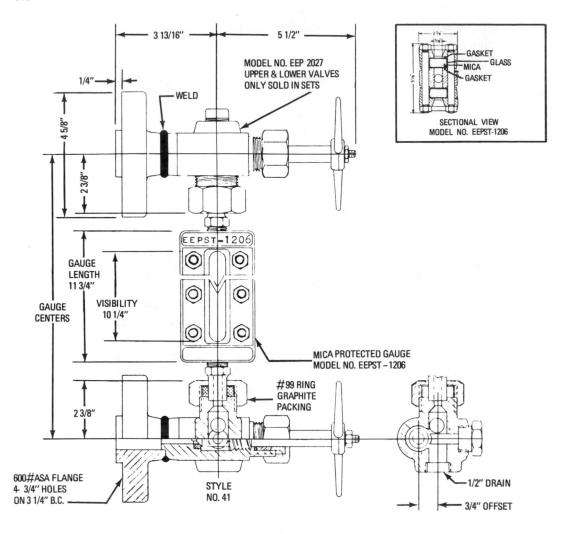

Figure 5-7. Flanged boiler navy water gauge. *(Courtesy of Eugene Ernst Products)*

The main stop valve is usually a nonreturn globe type and is designed to prevent backflow of steam into a boiler when multiple boiler installations are all feeding into a common header (see Fig. 5-10). When the pressure in the boiler falls below the main pressure, the valve will close and remain closed until the pressure in the boiler is sufficient to open the valve. This function is advantageous when adding or removing boilers from service on multiple installations, since it is unnecessary to get on top of the boiler to open or close the main stop valve (see Fig. 5-11).

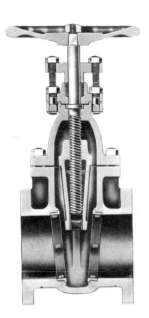

Figure 5-12. Section through a solid wedge gate valve with nonrising stem. All iron suitable for 200 psi W.O.G. *(Courtesy of Jenkins Bros.)*

Figure 5-13. Outside view of O.S. & Y. gate valve. The outside screw protects stem threads from attack by line fluids. *(Courtesy of Jenkins Bros.)*

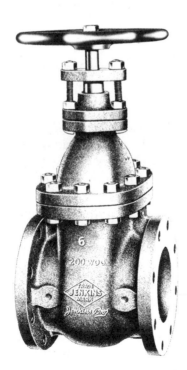

Figure 5-14. Section through O.S. & Y. solid wedge gate valve with renewable seat rings. Suitable for 125 psi steam. *(Courtesy of Jenkins Bros.)*

Figure 5-15. Outside view of inside screw nonrising stem gate valve. *(Courtesy of Jenkins Bros.)*

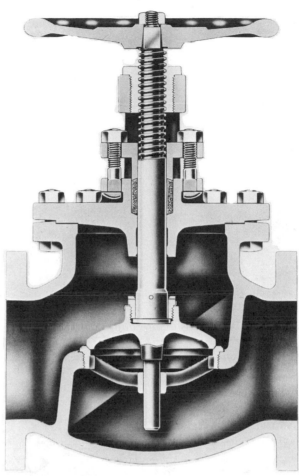

Figure 5-8. Section through O.S. & Y. globe valve with regrinding bronze disk and seat ring. *(Courtesy of Jenkins Bros.)*

It is normal practice to install globe valves with the pressure under the seat. This allows for repacking of the valve stem when the valve is closed and assists in opening the valve. There are installations such as those that use high-temperature steam where the valve should be installed with the pressure above the seat. This is a safety feature: the valve will automatically close if the valve stem should break. The particular type used for main boiler stop service is usually a "Y" pattern nonreturn stop/check valve with an outside screw and yoke as illustrated. The "Y" pattern reduces the pressure drop through the valve.

The gate valve, as its name implies, is literally a gate in the pipe

Figure 5-9. Outside view of O.S. & Y. globe valve. A control valve for noncorrosive service. *(Courtesy of Jenkins Bros.)*

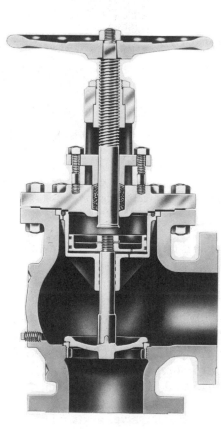

Figure 5-10. Section through non-return angle stop/check valve. Note bleed port in housing to piston. Valve will open when boiler pressure is greater than line pressure. *(Courtesy of Jenkins Bros.)*

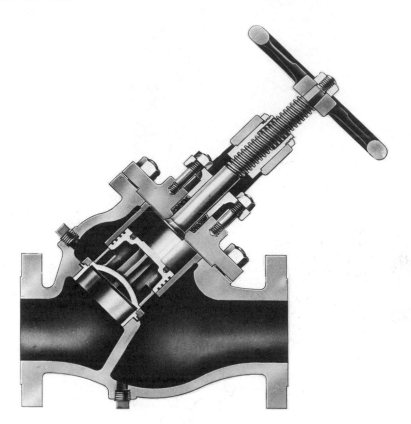

Figure 5-11. Section through Y-pattern globe valve nonreturn stop/check O.S. & Y. (*Courtesy of Jenkins Bros.*)

line (see Figs. 5-12 and 5-13). When the hand wheel is turned to open, the gate rises, opening the valve. Fluid or vapor flows freely through the valve in full open position with very little pressure drop or restriction. The valve will handle any fluid or vapor at a wide range of temperature, but it must be remembered that the valve can only be wide open or closed. It is not a throttling valve and its use in throttling service, with the gate partly open, will result in vibration and rapid wear of the gate and seat. Large gate valves are sometimes used on low-pressure boilers as the main stop valve and on main headers on multiple boiler installations.

An essential addition to any large gate valve used in steam service is the installation of a bypass around the gate. The gate valve has thickened sections called bosses cast into the housing. These bosses should be drilled and tapped to accommodate the bypass. When used

in steam service, the balancing of the pressure on either side of the gate will make opening the gate very easy and will also prevent water hammer in adjacent piping.

It is usual to select a gate valve for steam service with an outside screw and yoke (O.S. & Y.) with a rising stem (Fig. 5-14). The rising stem makes it apparent from a distance whether the valve is open or closed and also protects the threads on the stem from exposure to the flow of steam. To illustrate the difference, an inside screw gate valve with nonrising stem is shown in Fig. 5-15.

On both gate and globe valves the stem packing should be kept in good condition and not be permitted to leak. You should lubricate stem threads at regular intervals on larger valves. Grease fittings are installed to facilitate this lubrication.

When opening and closing both types of valves, beware of the expansive properties of the metals used in construction of the valve. If you close a gate or globe valve tightly when the valve is cold and later let steam flow into the system with the valve still closed, the expansion of the stem will increase the tightness of the valve until it is impossible to open. A valve that is cold should be closed only "hand tight." Manufacturers design the control wheel on the valve so that only the hand should be needed to close a valve "leak tight" (see Fig. 5-16). The use of a wrench or other device should never be necessary to close a valve. If a valve cannot stop the flow it obviously has a worn seat or is itself worn, and should be repaired or replaced. When a valve is opened cold, always ease back on the wheel a quarter turn to prevent the valve jamming in the open position because of its previously noted tendency to expand.

The forerunner of the valve was the cock, a simple device used for centuries and still particularly applicable for low pressure gas service and gauge shutoff. The plug cock should be used only as is the gate valve, for on–off service. Throttling will cause excessive wear. The plug cock and variations of it are also used in multidirectional flow service such as on fuel oil strainers when it is necessary to shift from one set of strainers to another.

A modern version of the centuries-old plug cock is the ball valve (Figs. 5-17 and 5-18). This valve gives little pressure drop since it has a straight-through port that permits unobstructed flow. Throttling will cause excessive wear. The modern ball valve improved on the plug cock, reducing the turning torque by having no metal-to-metal contact. The factor that made development of the modern ball valve possible was the availability of sealing and seat material. Therefore,

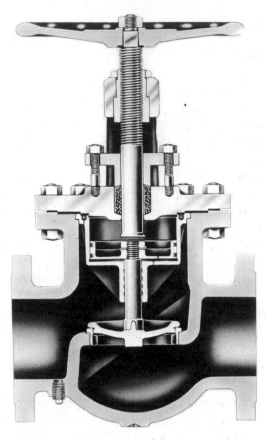

Figure 5-16. Section through nonreturn stop/check with O. S. & Y. *(Courtesy of Jenkins Bros.)*

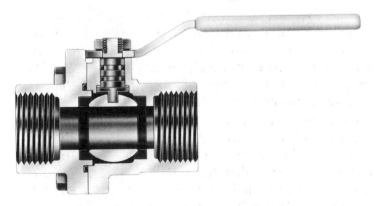

Figure 5-17. Section through threaded ball valve (note full bore port). *(Courtesy of Jenkins Bros.)*

Figure 5-18. Threaded ball valve. *(Courtesy of Jenkins Bros.)*

the most important factor in using these valves is selecting the correct materials for the intended service. The most common sealing material for use in steam and chemical systems is Teflon. This material can be used with the correct metal up to 400°F and 600 psi.

When is a valve not a valve? When it's a check valve. The check valve is used for controlling the direction of flow of liquids and gases and preventing backflow from high-pressure lines to lower-pressure lines. Horizontal swing check valves are normally used with gate valves, as they also ensure almost unrestricted free flow (Figs. 5-19 and 5-20). Fluid velocities must be low and nonpulsating; otherwise the check will constantly strike the seating surface and destroy it. The swing check valve should be installed only in horizontal or vertical lines where the flow-lift checks are normally used with globe valves since they have similar flow characteristics. To prevent backflow on steam, air, water, or gas lines with high flow velocities, horizontal lift checks must be installed on horizontal lines with the pressure under the seat. Vertical lift check valves can be installed only in a vertical position and must be installed with internal parts completely vertical. Any deviation from the vertical will cause the valve to stick in the open or closed position. In every boiler operation a check valve is required on boiler makeup service to prevent the pressure in the boiler from flowing into the feed water lines.

As with all other equipment in the boiler room, proper maintenance and operation practices will result in long, trouble-free life for boiler room valves.

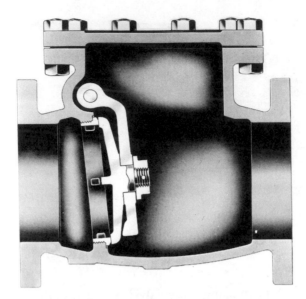

Figure 5-19. Section through horizontal swing check valve. Suitable for nonreturn service on steam, condensate, air, oil, and boiler feed lines. *(Courtesy of Jenkins Bros.)*

Figure 5-20. Outside view of horizontal swing check valve with bolted cover. *(Courtesy of Jenkins Bros.)*

SOOT BLOWERS

The ability of a boiler to maintain efficient operation is dependent to a great degree on the unimpeded transfer of heat from the fire side to the water side of the boiler. The necessity to limit the formation of scale on the water side has its counterpart in the need to keep the fire side free of deposits of soot and ash, the unburned or unburnable products of combustion. Since soot has five times the insulating quality of asbestos, a very minor deposit of soot on boiler tubes will result in a major decrease in operational efficiency. A ⅛-in. deposit of soot can result in a 13% increase in fuel consumption for the same boiler load.

The periodic shutdown of the boiler and the lancing of the fire tubes to remove accumulated soot deposits was common practice until the invention and adaptation of soot blowing lances. One of the earliest soot blowers was that produced by the Diamond Blower Company, which has been making them for 70 years.

All soot blowers have basically the same purpose: to remove deposits from the tubes by means of a dry blowing medium. The most common medium has been steam, but recent designs have made use of air and water. Water is used only in special circumstances where the use of high fouling fuel is an economic necessity but causes a buildup on the fire side of heavy noncombustible deposits. The use of water on the fire side of any boiler is not recommended. If applied while the boiler is steaming, the thermal shock to the tubes results in tube damage and if applied when the box is cold the formation of corrosive acids results in fire-side corrosion.

The selection of the type of soot blower depends upon the type of boiler and the fuel burned. There are various soot blowers available but (see Fig. 5-21) regardless of the type used, certain operating characteristics and safety precautions apply to them all:

(1) The boiler should be on manual operation with a stabilized flame.
(2) The velocity of air flow through the boiler must be sufficient to sweep all loosened soot and noncombustibles clear of the boiler.
(3) The steam should be as dry as possible and at full blowing pressure. Drains from the supply header should be left in the "cracked" open position when blowing to remove any condensate.

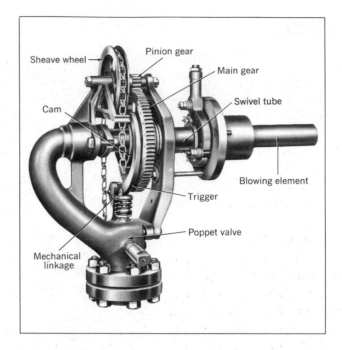

Figure 5-21. Diamond G-9 soot blower. *(Courtesy of Bab-cock & Wilcox)*

   (4) Once blowing has started do not leave blowing station; com-plete blowing in a continuous revolution.

   (5) If a smoke detector is in the stack, clean the lens after blow-ing tubes.

BOILER BLOWDOWN

The blowdown valves on a boiler are used for the following purposes:

   (1) Draining the boiler.

   (2) Lowering the water level.

   (3) Removing excess chemicals and precipitated sludge from the boiler water.

    For operational purposes, the need to drain the boiler prior to its cleaning or repair is self-evident. Also, when the water level rises to unacceptable levels, the necessity to lower the water level to prevent priming or water carryover into the steam line is also apparent. Item

3 cited above is the most important use for the blowdown valves when the boiler is operational and is an integral part of the water treatment for any boiler. The most important part of blowing a boiler down is to know exactly why you are blowing it down, because of the consequent loss of heat and therefore boiler efficiency when the boiler is blown down.

It is always necessary to blow a boiler down into a blowdown tank. The injection, into the local sewer system, of high-pressure water is prohibited by most municipalities so dissipation of the pressure and cooling of the contents of the boiler is necessary. All boilers should therefore be blown into a tank designed as a pressure vessel with an overflow to a sewer system, using a method of cooling the water before it reaches the sewer. Figure 5-22 illustrates a typical blowdown tank that can be used for spasmodic or continuous blowdowns.

All blowdowns result in a loss of heat, but this heat can be recovered with the installation of a heat-recovery unit such as the one illustrated in Fig. 5-23. This type of unit is best used with a continuous blowdown system piped as shown in the figure. This type of

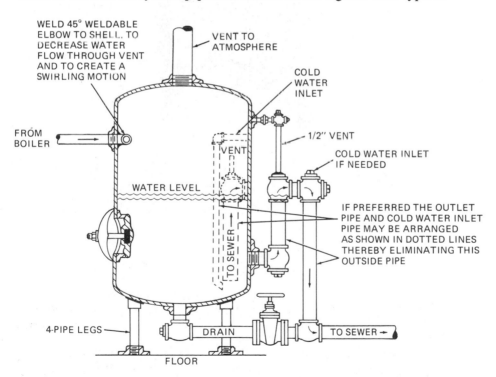

Figure 5-22. General arrangement blowdown tank.

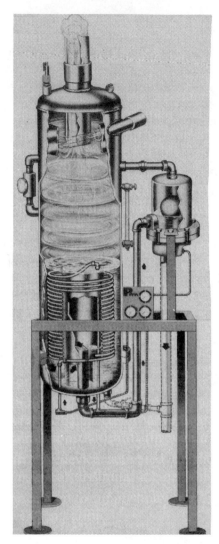

Figure 5-23. Blowdown heat recovery unit. *(Courtesy of Penn Separator Corp.)*

system combines both the intermittent blowdown and the continuous blowdown, with an ability to recover the heat when using a continuous system.

If we are operating a boiler at 150 psig, the blowdown heat loss must be calculated from the temperature of the steam in the drum compared to the temperature of the makeup water, say 50°F. Therefore blowdown loss in such a boiler could be:

Water

Enthalpy of water at 366 °F = 338.54 Btu's/lb

Enthalpy of water at 50 °F =   18.07 Btu's/lb

Subtracting the difference = 320.47 Btu's/lb

This blowdown loss would equal 320.47 Btu's for every pound of water blown out.

BURNERS

Fuel oil burners are divided into three broad categories: mechanical atomization, steam- or air-assisted atomization, and centrifugal atomization.

Mechanical atomization depends solely upon the pressure of the oil at the burner, as it is forced through a specially designed series of holes and grooves to achieve atomization, or the breaking up of the oil stream into small particles. This has the effect of presenting to the oxygen, in the combustion air, a larger surface area with which to combine with the hydrocarbons in the oil.

Mechanical atomization, when used with the lighter fuel oils, such as No. 2, provides acceptable atomization without the need to preheat the oil. Heavier fuels, such as No. 4 and No. 6 oil, require heating to reduce the viscosity. The temperature to which the fuel has to be raised in order to burn efficiently is subject to trial and error, and the optimum fire point of the fuel is variable.

Steam or air-assisted atomization is possible with much lower oil pressures at the burner (Figs. 5-24 and 5-25). The steam or air combines with the oil in the nozzle of the burner, breaking up the oil into particles. Heavy oils require preheating to reduce the viscosity and when steam-assisted atomization is used, heating the oil prevents condensation of the steam in the burner barrel.

Centrifugal atomization is achieved with a rotary or cup burner. The atomizing cup is rotated at high speed, by a direct or belt-drive motor. The cup is smoothly tapered and oil is fed into the small end of the cup. As the cup rotates, oil, in a thin film, moves forward and flows off the edge of the cup. A high-velocity air stream, directed by vanes into a direction opposite to the direction of rotation of the cup, breaks the film of oil leaving the cup into fine particles and propels it forward into the furnace.

Figure 5-24. CN type burner. *(Courtesy of Cleaver Brooks)*

All burners require an initial form of ignition, such as an electric spark, gas flame, or a hand-held torch. Once started the flame will be self-sustaining, since the heat from the oil already burning causes vaporization and ignition of the following oil.

An alternative fuel to oil is natural gas, and in some installations oil and gas are burned in combination. Unlike oil, gas requires no atomization being already finely divided.

Two types of burners can be used, a nozzle mix burner and a premix burner. In the nozzle mix burner the mixing of the gas with the combustion air takes place in the furnace, where the gas entering

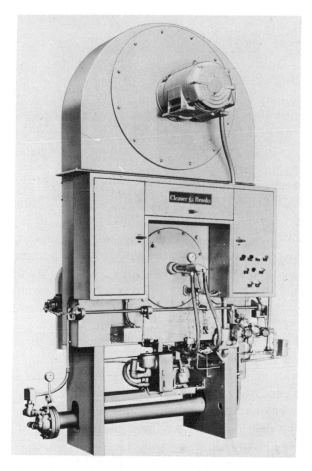

Figure 5-25. BR burner. *(Courtesy of Cleaver Brooks)*

the furnace passes through a gas ring set into the refractory at the furnace front. Premix burners, used on smaller heating units, utilize a venturi nozzle through which the combustion air flows inducing the gas into the air stream and ensuring adequate mixing prior to combustion.

All burners require control. On hot-water boilers the controlling factor is the temperature of the water and on steam boilers steam pressure is utilized for on–off control. If the boiler supplies large quantities of steam at varying loads, continual monitoring of the steam flow is used to vary the flow of oil or gas. Regardless of the fuel used, constant attention to flame quality is necessary on the part of the operator to ensure efficient operation.

COAL BURNING

Oil and gas superceded coal as a boiler fuel, but with the widely publicized finite limit on the supply of fuel oils and gases, the need to burn coal may again become a necessity.

In the early days of coal firing, the coal was fired by hand. The skill needed to coal, slice, and rake a boiler furnace without losing steam pressure has been replaced with automatic stokers and pulverized fuel burning.

The stokers are made in various forms, depending on boiler size and fuel used. These include travelling grates, vibrating grates, and retort stokers. When burning wastes and garbage in incinerators or using the waste to generate steam, specially designed moving grates are needed to continually aerate the garbage. The dumping of ashes is accomplished automatically with various designs of dumping grates.

The stoker illustrated in Fig. 5-26 is a hydraulic ram feed stoker with side dumping. Coal is fed by the hydraulic ram from the hopper and tunnel section into the retort, onto the tuyeres and live plates. Combustion air from under the stoker is distributed through the holes in the tuyeres and live plates, to provide even burning. An overfire air system with manifold, air jets, blower, and control valve ensures smokeless combustion.

The majority of boilers burn a single fuel—coal, gas, or fuel oil. Some have the ability to burn two fuels, either singly or simultaneously.

The boiler illustrated in Fig. 5-27 is a tri-fuel boiler, burning coal, oil, and gas. This boiler is designed with the ability to burn coal, oil, or gas separately or fire coal with oil or gas.

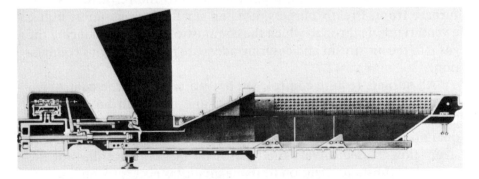

Figure 5-26. Hydraulic ram feed stoker-side dump. *(Courtesy of North American Manufacturing Co.)*

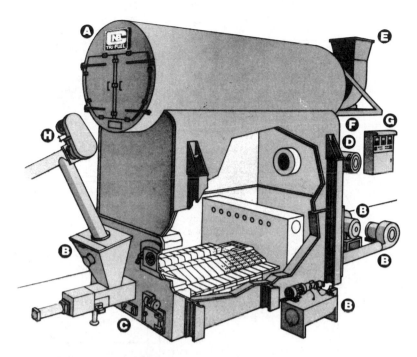

Figure 5-27. CNB tri-fuel boiler. *(Courtesy of North American Manufacturing Co.)*

The ability to fire either gas or oil with coal reduces particulate and SO_2 emissions. It permits operation with a constant coal burning rate and modulates gas or oil to meet varying loads.

The boiler is a fire tube, fire box boiler of three-pass construction. The stoker is a simple retort, side dump, under feed type hydraulically operated. The dual fuel burner is an air atomizing oil burner and the gas burner is capable of burning any gaseous fuel.

This combination of burning equipment provides the operator with the necessary flexibility to burn different fuels at economical rates.

FEEDWATER SYSTEMS

The addition of water to a boiler when producing steam is a continuous process, requiring reliable equipment and control. An example of a feedwater system is shown in Fig. 5-28.

Modern practice makes use of the centrifugal pump to raise the pressure of the water above that of the pressure in the boiler in order

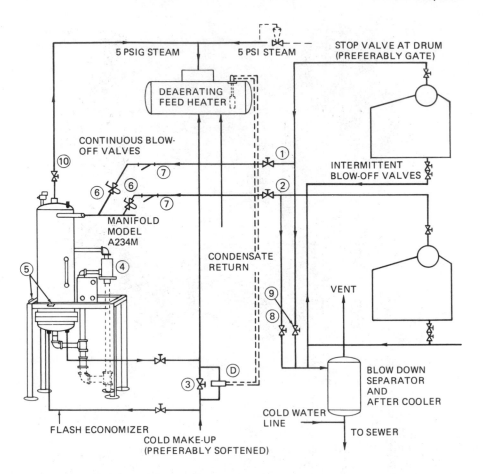

Figure 5-28. Feedwater system. *(Courtesy of Penn Separator Corp.)*

to force the water into the boiler. The motive power is provided by an electric motor or a steam-driven turbine. In some older plants, direct-acting duplex steam-driven pumps are still in service but for economy of operation the centrifugal electric-driven pump is preferred (Fig. 5-29).

In determining the quantity of feedwater needed to be delivered to the boiler when the boiler is operating at full capacity, the A.S.M.E. standard of 1 boiler horsepower being equivalent to 34.5 lb of water at 212 °F is used.

To find out the number of gallons of water needed per hour, the rated horsepower of the boiler is multiplied by 34.5 and divided by the weight of 1 gallon of water.

Figure 5-29. Spraymaster deaerator. *(Courtesy of Cleaver Books.)*

Example

Assume a boiler is rated by the manufacturer at 300 hp. How many gallons of water must be delivered to the boiler when operating at full capacity?

$$300 \times 34.5 = 10,350$$

Assume 8 lb/gal as the weight of the water.

$$\frac{10,350}{8} = 1293.75 \text{ gal}$$

Therefore, to provide the necessary water, the pump must theoretically deliver 1293.75 gal/hr. In actual practice, to compensate for pump wear and the possibility of the pump delivering to a height, this value is increased by 30% or use 45 lb of water per boiler horsepower.

The feed piping and valves to the boiler must be adequately sized to allow free flow of water without restriction when the boiler is producing steam at rated capacity.

Immediately next to the boiler, a stop valve must be placed in order to isolate the entire feed system from the boiler pressure, if needed. Next to the stop valve, a check valve is located, to ensure that the flow of water is in one direction only, into the boiler.

The valve manifold around the level control valve should consist of a stop valve on either side of the level control valve and a bypass valve.

The sizing of the control valve and associated valves to provide a low pressure drop is important. It is equally important to size the control valve small enough so that when the valve is operating to provide feed water it will have to open at least 50% of its travel. A too large control valve that is required to open slightly to allow water into the boiler will result in wire-drawing of the valve and valve seat and consequently a leaking control valve.

Another method of feeding a boiler with water is by means of an injector. While not used to any great extent in modern practice, it is installed on some installations as an emergency system. Fed with cold water and using steam as the propelling force, an injector can be used to force water into a boiler, the steam supplying heat to the cold water. Most injection systems are considered very tempermental and will only work if all design criteria are met, but in an emergency situation they can prove of value.

EXERCISES

1. A boiler operates at 120 psig. The safety valve is 6 in. in diameter. What is the load on the valve?

    Ans.:   3393 lb

2. The load on a safety valve is 2000 lb. If the valve is 4 in. in diameter, at what pressure will the safety valve lift?

    Ans.:   159.15 lb/in^2

3. What is the blowdown on a safety valve? Describe its purpose.

4. A safety valve is set to lift at 300 psig. The spring load is 1500 lb. What is the valve area?

    Ans.:   5 in.

5. In Question 4 when the valve lifts the exposed area increases to 5½ in². At what pressure will the valve reseat itself?

    Ans.: 272.72 psig

6. A boiler has a total heating surface of 3500 ft². The safe working pressure is 250 psig. If the specific volume of the steam is 1.75, what area of opening is necessary for the safety valves?

    Ans.: 14.58 in²

7. What would be the total heating surface on a boiler operating at 100 psig, if the safety valve opening is 6 in² and the specific volume of the steam is 4.2?

    Ans.: 600 ft²

8. You entered the boiler room and blow the gauge column down and water did not appear in the glass. What would be the cause?

9. Describe an O.S. & Y valve.

10. Why do you use a soot blower in the boiler room?

11. State all the reasons for blowing a boiler down.

12. A boiler is rated at 1000 bhp. When developing its full capacity, how many gallons of water will be delivered by the feed water pumps per hour? Assume water weighs 8 lb per gallon.

    Ans.: 4,312.5 gallons

13. What is the purpose for atomizing oil?

14. Describe a hydraulically driven ram feed stoker.

MULTIPLE CHOICE QUESTIONS

1. An accumulation test is given to determine:
   a. Furnace volume
   b. Safety valve capacity
   c. At what pressure the safety valves will lift

2. The most important valve on a boiler is the:
   a. Safety valve
   b. Blowdown valve
   c. Feedwater valve

3. The first duty of an operator when starting his shift is to:
   a. Check the water
   b. Check the fuel
   c. Check the fire

4. A globe valve is usually used to:
   a. Check flow
   b. Throttle flow
   c. Stop flow

5. On low-pressure boilers the main stop valve is usually:
   a. A gate
   b. A globe
   c. O. S. & Y.

6. Soot on tubes will result in:
   a. Loss of efficiency
   b. Loss of steam
   c. Loss of water

7. Blowdown lines should never be connected directly to:
   a. Flash tanks
   b. Steam lines
   c. Sewer lines

8. Fuel oil mechanical atomization depends on:
   a. Viscosity
   b. Pressure
   c. Temperature

9. One boiler horsepower equals:
   a. 34.5 gal/hr
   b. 34.5 lb/hr
   c. 34.5 lb/min

10. Feedwater lines to a boiler must always be equipped with:
    a. An O. S. & Y. valve
    b. A gate valve
    c. A check valve

6

Boiler Auxiliaries

PUMPS AND PUMPING

One of the most important pieces of auxiliary equipment in any boiler room is the feedwater pump, (see Fig. 6-1). Except for those, installations where boiler pressures are low and feedwater to the boiler comes directly from a low-pressure water source, all boilers are fed via a pump. Failure of this pump will result in a shutdown of the boiler. In installations that burn fuel oil, failure of the fuel oil pump will also cause plant shutdown.

Pumps for use in the boiler room are generally of three types: positive displacement, centrifugal, or rotary.

A positive displacement pump is one that comprises a piston within a cylinder; the movement of the piston within the cylinder displaces the liquid, forcing it through a valving system. Positive displacement pumps can be single-acting and take in liquid on only one side of the piston, or double-acting when both faces of the piston are utilized. A further refinement is two pistons and cylinders, called a duplex pump, or three pistons and cylinders, called a triplex pump. Most positive displacement pumps of this type are steam-driven with a reciprocating engine.

The centrifugal pump is a pump in which the discharge pressure is created by the imposition of centrifugal force to the liquid being

Figure 6-1. GT 2 stage high-pressure pump suitable for feed-water service. *(Courtesy of Ingersoll-Rand Co.)*

pumped (see Fig. 6-2). Centrifugal pumps are generally electric-motor or steam-turbine driven and can be classified according to the design of the impeller and the number of impellers or stages. For example, a single impeller of open design would be classified as a single-stage open-impeller centrifugal pump. Similarly, a centrifugal pump with two or more impellers in the same casing is classified as a multistage pump: two-stage, four-stage, or ten-stage, depending on the number of impellers used. The number of stages is theoretically limitless. Various examples of centrifugal pumps are shown in Figs. 6-3 through 6-7.

A rotary pump is a positive displacement pump with a casing containing a rotary element consisting of screws, gears, valves, or some modification of these elements. Most types of rotary pumps are suitable for handling viscous liquids such as oils.

Each of the pump types cited is classified by the kind of force imparted to what is being pumped, after it has reached the pump. A

Figure 6-2. Typical installation 2-stage high pressure boiler feed pump. *(Courtesy of Ingersoll-Rand Co.)*

Figure 6-3. Large single-stage, double-suction horizontally split case pump. *(Courtesy of Ingersoll-Rand Co.)*

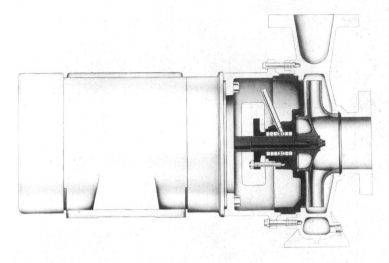

Figure 6-4. Close-coupled centrifugal pump. *(Courtesy of Ingersoll-Rand Co.)*

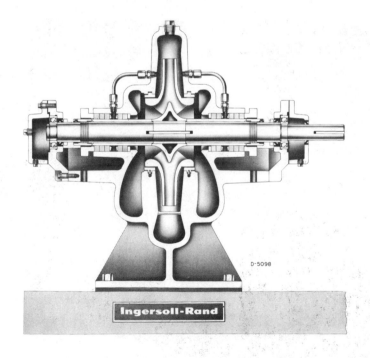

Figure 6-5. Horizontally split case centrifugal pump. *(Courtesy of Ingersoll-Rand Co.)*

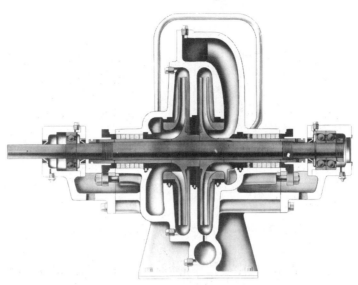

Figure 6-6. Cross-section view of a high-pressure boiler feed pump. *(Courtesy of Ingersoll-Rand Co.)*

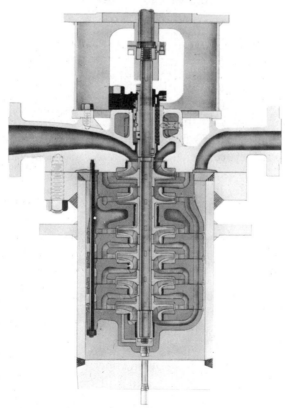

Figure 6-7. Vertical multistage diffuser pump. *(Courtesy of Ingersoll-Rand Co.)*

common misconception is that a pump "pulls" what is being pumped unassisted by any outside force. This is not true. The pump actually creates a low-pressure void and the pressure differential between the pump inlet low pressure void and the external forces applied to the liquid causes it to flow into the void, or suction side of the pump.

This external force is called the Net Positive Suction Head (N.P.S.H.) and is the sum of all the plus or minus forces acting upon the liquid and forcing it into the low pressure void created by the pump. These forces are:

(1) Atmospheric pressure: Normally 14.7 psia or 30 in. of Mercury (Hg).
(2) Plus. Static head in psi (level of liquid above pump inlet).
(3) Minus. Static lift in psi (level of fluid below pump inlet).
(4) Minus. Inlet line friction losses.
(5) Minus. Fluid vapor pressure.

Calculating the five forces yields the net suction head. Assuming that a pump is in good mechanical condition and with a leak-free suction line, the height to which water may be drawn by a pump depends upon the friction in the suction pipe, the temperature of the water and altitude above sea level. Atmospheric pressure at sea level is approximately 14.7 lb/in² absolute. A column of water of any cross-sectional area 2.31 ft in height will provide a pressure of 1 lb/in² at its base; therefore the theoretical suction lift at sea level is 14.7 × 2.31 = 33.957, or about 34 ft. If friction losses through the pipe are taken into consideration, plus the temperature of the water, this theoretical figure will be—as calculated in the days of Galileo and Leonardo da Vinci—reduced in actual practice to 24 to 26 ft.

The temperature of the water being pumped has a profound effect upon the ability of the pump to move the water. As noted previously, the creation of the zone of low pressure by the suction of the pump will permit water at an elevated temperature to flash to steam, causing the pump to become vapor bound and inhibiting its ability to maintain the low-pressure zone necessary to move liquid. In all boiler rooms of good design, the feed heater is situated at the highest point in the boiler room to allow the operator to raise his feedwater temperature as high as possible and still be able to pump without a vapor binding of the pump. Boiler feed pumps drawing water from a zero-pressure feedwater heater should have about 2 ft static head for water at 165 °F. A column of water 1 in. square and 1 foot high weighs

0.433 lb. Thus at 11.5 ft the pressure at the suction of the pump would be 4.979 lb/in². Water at this pressure and temperature will not vaporize. The pressure due to the height of a column of water may be converted into pounds per square inch by use of the following formula:

$$P = \frac{H}{2.31}$$

where P is pressure in pounds per square inch and H is the height, or head, in feet.

Example

A feed heater is situated on an upper platform in a boiler room. The distance from the surface of the water in the feed heater to the pump suction is 34 feet. What is the pressure at the suction of the pump?

$$P = \frac{H}{2.31} = \frac{34}{2.31} = 14.7 \text{ lb/in}^2$$

All pumps must overcome certain restrictions to flow. These are the velocity head, the friction head, and the measured head.

The velocity head is the force required to set the liquid in motion. The friction head is the force required to overcome the friction between the liquid and the walls of the piping and internal parts of the pump. The measured head is the vertical distance between the pump suction and the lower or higher plane from which the liquid is drawn plus the vertical distance to the point of discharge plus any internal pressure to be overcome at the point of discharge.

When specifying a pump for boiler-room use, keep in mind that unless the piping system is extensive, the pressure losses in the velocity head and friction head are of such small amounts that they are often neglected and indeed can be in most cases. The factors affecting both the velocity and friction heads are the number of fittings (elbows and valves) in the system and the smoothness of the pipe.

For all practical purposes, when pumping water, the total dynamic head, which is the sum of the total discharge head and the total suction lift (see Fig. 6-8) will be sufficient to size a pump.

When pumping high-viscosity liquids it is necessary to take the level of viscosity into account. It is the determining factor affecting frictional head, motor size required, and speed of operation necessary. When pumping oils or other volatile elements, the vapor pressure of the liquid is also a consideration.

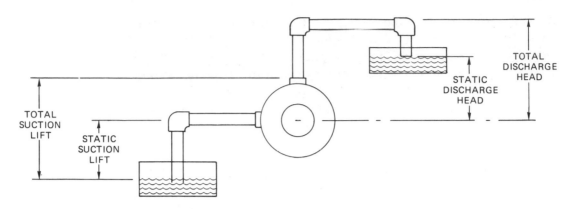

Figure 6-8. Total discharge head + Total suction head =
Total dynamic head.

All liquids boil at a specific temperature and vaporize at certain
levels of pressure and temperature. As the pressure is reduced, boiling
will occur at a lower temperature. Water boils at atmospheric pressure
at sea level at 212 °F. As boiling takes place vapor is given off by the
liquid.

For most common liquids at ambient temperature, boiling occurs
below atmospheric pressure. As the pressure in the suction line decreases, the vacuum increases and a pressure is reached at which the
liquid boils. This pressure is known as the *vapor pressure* of the liquid.
As the action of the pump creates a low-pressure condition in the
suction line, both vapor and liquid will enter the suction of the pump
and the capacity of the pump will be reduced. When the vapor bubbles
enter the pressure or discharge side of the pump, the bubbles collapse, causing noise and vibration. The formation of vapor in the
suction line and the resultant collapse of the vapor bubbles in the
pump is called *cavitation.* Restricting the suction line of a boiler feed
pump and thus increasing the vacuum in the suction inlet to the pump
will result in cavitation.

The rate at which a pump does work may be expressed in terms
of horsepower. The work required to drive the pump or the power
input is designated as brake horsepower. Friction and slippage in a
pump are the main sources of loss of power so that the power output
is always less than the power input. Pump efficiency is defined as
power output divided by power input. Hence,

$$\text{Efficiency} = \frac{\text{Power output}}{\text{Power input}}$$

Example

A pump discharges the equivalent of 15 bhp. If the pump is driven by a 20 hp motor what is the efficiency?

$$E = \frac{\text{Power output}}{\text{Power input}}$$

$$= \frac{15}{20} = 75\%$$

Efficiency values for different pumps will vary widely depending upon the operating conditions and the design of the pump. The pumping efficiency will also vary depending upon the quantity of liquid being pumped: For example, a pump may deliver 100 gallons per minute at an efficiency of 40% and 300 gallons per minute at 70% efficiency. All manufacturers of pumps provide what are known as pump curves. These curves, on a graph, illustrate the capacity, efficiency, and brake horsepower of a specific design of pump at varying conditions of speed, impeller size, horsepower and head.

In order for any pumping to take place, the velocity that is imparted to the liquid being pumped, by the centrifugal force of the impeller, must be of sufficient force to overcome any restriction in the discharge pipe. If a ball is thrown into the air and rises to a height of 10 ft and then falls back into the hand that threw it into the air, the force required to throw it to a height of 10 ft will be the same force at which it will strike the hand when falling through a distance of 10 ft. The velocity which a body will acquire when falling is the same velocity that will cause it to rise to the height from which it fell. This velocity can be calculated with the formula:

$$V^2 = 2gs$$

or

$$V = \sqrt{2gs}$$

where V is velocity in feet per second

g is the acceleration of gravity 32.2 ft/sec/sec

s is the distance in feet through which a body falls

Example

What velocity in feet per second must be imparted to a weight to project it 120 ft in the air?

$$V = \sqrt{2gs}$$

$$= \sqrt{2 \times 32.2 \times 120}$$

$$= 87.9 \text{ ft/sec}$$

Note also that:

$$60 \times \sqrt{2 \times 32.2} = 481$$

Therefore:

$$\text{Velocity in feet per minute} = 481\sqrt{s}$$

Conversely, suppose a body falls from a height of 120 ft. What will its velocity be in feet per minute when it strikes the ground?

$$V = 481\sqrt{s}$$
$$= 481\sqrt{120}$$
$$= 5269 \text{ ft/min}$$

and

$$\frac{5269}{60} = 87.9 \text{ ft/sec}$$

By the same reasoning, the water to be pumped to a certain height must have the same velocity on leaving the pump as it would have if it fell through the same distance in feet or "head," and the head can be reduced to pounds per square inch by dividing by 2.31. Therefore the theoretical velocity at the periphery of a centrifugal impeller will govern the revolutions per minute that the pump has to be driven to supply sufficient velocity to the liquid being pumped to overcome the total head.

Example

At what r/min must a centrifugal pump with a 12-in.-diameter impeller be driven to deliver water against a total head of 120 ft?

$$V = 481\sqrt{120}$$

$$= 481 \times 10.95 = 5269 \text{ ft/min}$$

The circumference of the impeller is $\pi \times$ diameter. Therefore:

$$3.1416 \times 1 = 3.1416 \text{ ft}$$

Therefore

$$\frac{5269}{3.1416} = 1677 \text{ r/min}$$

Because of friction and other losses that cannot be entirely eliminated the revolutions per minute would be slightly greater in order to produce the given head.

The quantity of water delivered by a centrifugal pump will vary directly with the speed of the impeller, all other factors remaining constant. This can be expressed:

$$Q = \frac{R_1 \times Q_1}{R_2}$$

where Q is the quantity of water delivered in gallons per minute at R_1 revolutions per minute and Q_2 is the quantity of water delivered at R_2 revolutions per minute.

Example

A pump running at 1700 r/min delivers 600 gal/min. What would be the gallons per minute delivered at 850 r/min?

$$600 = \frac{1700 \times Q_2}{850}$$

$$600 \times 850 = Q_2$$

$$300 \text{ gal/min} = Q_2$$

There are various classifications of centrifugal pumps but the only type that can strictly be called a centrifugal pump is one in which the liquid enters the eye of the impeller and is caused to flow radially to the periphery by the action of centrifugal force.

In the volute type pump, the liquid entering the suction pipe is caught by the rotating vanes of the impeller and given a velocity which carries it through the spiral casing called the volute to the discharge nozzle. This volute is designed to produce an equal velocity flow around its circumference.

Water can be raised as high as necessary by arranging a series of pumps, one pump to take the discharge from another and thereby compound the discharge pressure.

PUMP PACKING

In any pump, except for the so-called "sealless pump," the point at which the rotating or reciprocating shaft enters the liquid end of the pump has to be sealed (see Fig. 6-9). This point is called the stuffing box or packing gland. Originally the name stuffing box was appropriate because this point was literally stuffed with a wide variety of materials to provide the necessary seal. In the earliest pumps, ropes of jute or flax were well greased to reduce friction and to prevent the packing from burning. In addition to greasing, the gland was allowed to leak water and thereby provide for cooling of the packing. Any

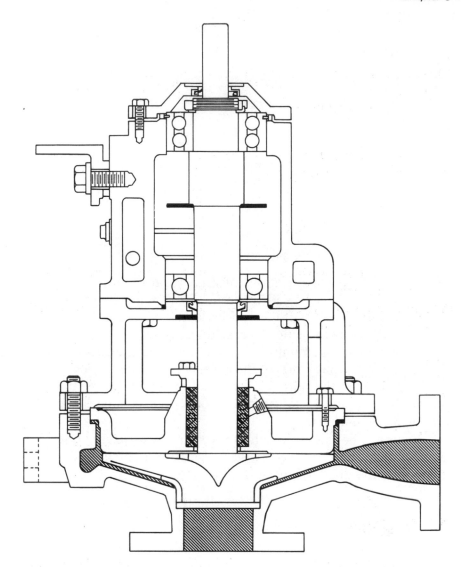

Figure 6-9. Packed pump. *(Courtesy of Sealol Inc.)*

packed pump will have a tendency to leak, and, indeed it is usual that
the gland drip slightly even when well packed in order to cool the gland.

In the years of experience gained in packing pumps, manufacturers
have developed many materials in different forms to develop packing
material that will perform the required function. To provide the
necessary sealing, a packing material must withstand friction from
the moving shaft, the fluid reaching it, the pressure from the pump
and gland and the temperature of the fluid being pumped. The most

common material found in boiler room packings are white and blue asbestos reinforced with rubber or lead. With recent advances in technology, plastics, graphite filaments, metallic compounds and combinations of these materials are now in common use.

The three main factors influencing the performance of any packing are:

(a) Quality of the packing. Use the best available, properly sized and designed for the service.
(b) Maintain equipment in good condition. A worn bearing or misaligned or worn shafts can reduce the life of packing substantially.
(c) Install packing correctly. The cost of labor to install packing incorrectly is the same as when the job is performed correctly the first time.

The correct method for packing a pump is as follows:

(1) Remove all old packing from the stuffing box. Clean the shaft and box thoroughly and examine the shaft for wear or score marks. Excessive wear or scoring of the shaft will reduce packing life substantially. If the shaft is badly worn it should be replaced.
(2) Use the correct size and type of packing to repack the box. To determine correct packing size, measure the diameter of the shaft and the inside diameter of the stuffing box. Subtract the shaft diameter from the stuffing box diameter and divide by two.

Example

A pump has a shaft diameter of 1½ in. The inside diameter of the stuffing box is 2½ in. What size packing is needed for this pump?

$$\frac{2\frac{1}{2} - 1\frac{1}{2}}{2} = \frac{1}{2}$$

Therefore, packing size would be ½ in. square packing.

It is probable that more pumps are packed incorrectly through use of incorrect packing size than for any other reason. This is the result of a lack of knowledge or laziness on the part of the person packing the pump. It is much easier to pack a pump with packing size smaller than the correct size. Standard coil packings are made in increments of $\frac{1}{16}$ in. and if the calculated size is slightly under a certain standard size use the next larger size.

(3) Cut the packing into rings. Never wind packing around the shaft and try to push it into the box. Cut the rings with a square joint. The best way to cut packing rings is on a shaft or pipe the same size as the shaft. Hold the packing tightly around the shaft or pipe and cut to the exact length needed. Do not stretch the packing. Cut with a butt joint, allowing no gaps. Try the ring in the stuffing box, making sure that it fits the space properly. Each successive ring can be cut in the same manner or the first ring can be used as the master ring to cut the other rings. When using the latter method be sure to cut the rings matched against the side of the master ring and not the O.D. or I.D. of the master ring.

(4) Install one ring at a time, making sure that no grit or dirt is on the ring. Oil or grease lubrication will assist when packing the ring into the box. The joints of each ring should be staggered, so that each joint is at least 90° away from the preceding joint. Each individual ring should be tamped home before putting in the next ring. When sufficient rings have been added so that the gland will reach them, supplement the tamping with use of the gland.

(5) When all packing has been installed, tighten the gland bolts all the way with a wrench to seat the rings of packing firmly. Then loosen the nuts of the gland studs until nuts are finger tight. Newly installed pump packing should be allowed to leak freely on startup. Take up gradually on gland nuts until only 2 to 3 drops per second is leaking. Never try to stop leakage entirely. As we pointed out earlier, all packed pumps leak, and the result of a nonleaking gland will be burning of the packing.

(6) If the pump is equipped with a lantern ring, to provide either grease or fluid lubrication, always be sure that the lantern ring is set slightly behind the hole provided in the stuffing box for such lubrication. When the gland is taken up, the lantern ring will move forward and if not positioned correctly the ring of packing immediately behind the lantern ring will blank off the inlet hole the first time the packing is tightened.

SEALING PUMPS

We have seen that, in order for a packed pump to run correctly, some slight leakage at the gland is required to cool the packing. Unless extraordinary measures are used to reclaim this leakage, it represents a loss in water and heat. The usual method is to allow this water and heat to run into a sewer. When the pump is used for boiler feed, the plant is losing chemically treated water that has to be made up by adding water to the system, requiring chemical treatment of the raw water. In any power plant, leakage means a loss of efficiency and higher cost of operation. To overcome this problem, the mechanical seal with zero leakage is rapidly becoming the standard in a well-run power plant.

The development of the mechanical seal over the last 20 years has been rapid, since many manufacturers have designed and modified seals to overcome deficiencies in performance. The multiplicity of mechanical seals available today makes the selection of the correct seal for a particular application very easy, but remember that the right seal for any use is the one that works.

The primary aim of any mechanical seal is to stop leakage; this is accomplished by opposing the fluid being pumped by two rubbing faces. These two faces, called the *seat* and the *seal,* provide the main seal to prevent the pressure in the pump from leaking to the atmosphere. The two rubbing faces must be compatible to provide as little friction as possible. A good selection for boiler room feed water is carbon against ceramic, although other materials may be used.

In addition to providing the seal surfaces, the seal must also resist the internal pressure being generated by the pump. In the case of a packed pump, it is the internal pressure developed by the pump that has to be contained. This is accomplished by placing a spring load on the seal to hold it on its seat with a force designed to prevent the internal pressure of the pump from pushing the seal off its face. The design of the spring or springs is unique to the use of the seal and the manufacturer's preference.

Examples of types of springs used, are illustrated in Fig. 6-10. The compressive force of the spring holds the seal on its seat to resist the pressure of the pump. Such is the case for a seal that is mounted outside of the pump stuffing box. When the seal is mounted internally, the pressure of the spring plus the internal pump pressure holds the

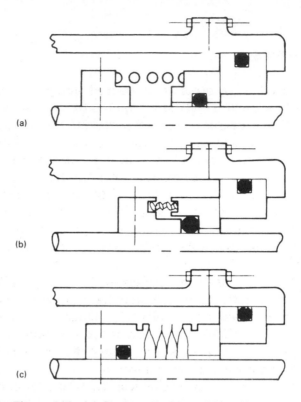

Figure 6-10. (a) Single coil spring; (b) multiple coil springs; (c) bellows. *(Courtesy of Sealol Inc.)*

seal on its seat. As the pump pressure increases, the force holding the seal on its seat also increases. (See Fig. 6-11).

This increase in pressure may be sufficient to overcome the spring pressure that is holding the seal on its face in an outside mounted seal. Where the internal pump pressure exceeds 50 psig it is usual to mount the seal internally so that the pump pressure will hold the seal on its face. Where pressures exceed 150 psig this face closing pressure may be too great, and excessive face closing pressure can damage the seal faces. In these instances it is necessary to use what is known as a "balanced seal." There is a lip on the face of the balanced seal that extends over a step in the shaft. Because the whole face area of the seal is not receiving the pump pressure, the seal is able to withstand the higher pump pressures.

In a mechanical seal, any leakage would occur across the rubbing faces, with all other avenues of leakage having been blocked by gaskets or packing. These packings can take the form of an O ring, a plastic bellows, or a V ring. The O ring is sized to provide a tight grip on the shaft, the bellows is clamped to the shaft, and the V ring

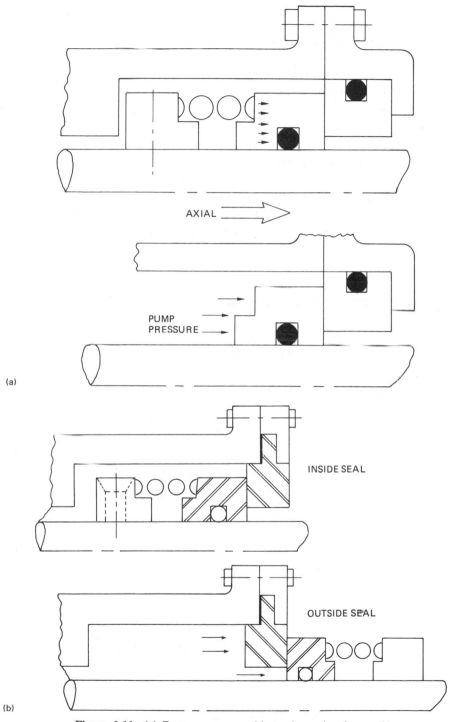

AXIAL

PUMP PRESSURE

(a)

INSIDE SEAL

OUTSIDE SEAL

(b)

Figure 6-11. (a) Pump pressure adds to the spring force; (b) inside and outside seals. *(Courtesy of Sealol Inc.)*

packing is placed so that any pressure from the pumped liquid tends to tighten it around the shaft.

If added sealing is needed because of pressure, temperature, or a dirty product being pumped, a double seal is sometimes used (Fig. 6-12). This seal provides two sealing faces to the fluid being pumped and an independent source of fluid to hold the seats on the faces. It is important to remember that, with this type of seal, the independent source of fluid must have sufficient pressure to hold the seal against the pump pressure. Cooling of the seal is accomplished using systems such as those illustrated in Fig. 6-13.

There is no standard procedure for installing a mechanical seal, but there are four critical requirements in any seal installation. These are:

(1) Determine that the pump is ready for a seal.
(2) Place the seal in the correct position for the right operating length.
(3) Prevent damage to the seal rings.
(4) Prevent damage to the seal faces.

Examples of installed seals are shown in Fig. 6-12

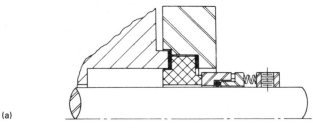

(a)

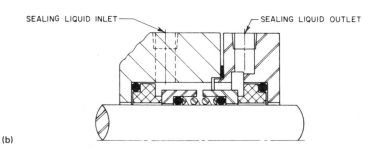

SEALING LIQUID INLET

SEALING LIQUID OUTLET

(b)

Figure 6-12. (a) Outside "RO" dura seal; (b) double "CRO" arrangement.

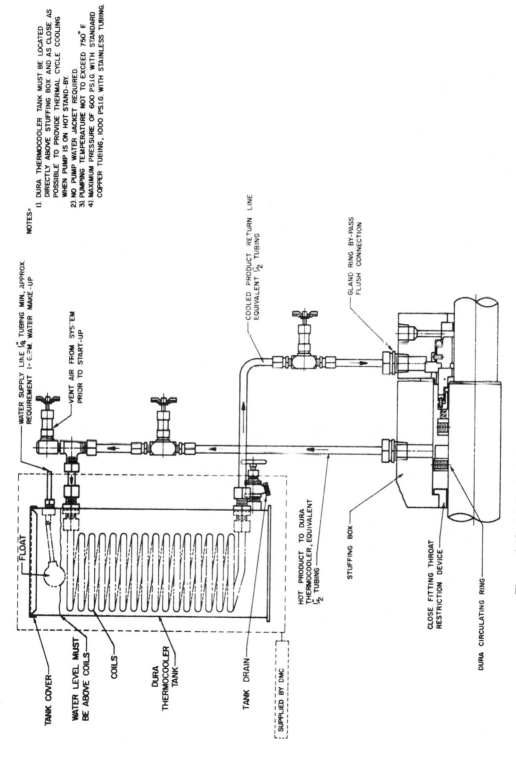

NOTES:

1). DURA THERMOCOOLER TANK MUST BE LOCATED DIRECTLY ABOVE STUFFING BOX AND AS CLOSE AS POSSIBLE TO PROVIDE THERMAL CYCLE COOLING WHEN PUMP IS ON HOT STAND-BY.
2) NO PUMP WATER JACKET REQUIRED.
3). PUMPING TEMPERATURE NOT TO EXCEED 750° F
4) MAXIMUM PRESSURE OF 600 PS.I.G. WITH STANDARD COPPER TUBING, 1000 PS.I.G. WITH STAINLESS TUBING.

WATER SUPPLY LINE ¼ TUBING MIN, APPROX. REQUIREMENT 1^ G.P.M. WATER MAKE-UP

VENT AIR FROM SYSTEM PRIOR TO START-UP

COOLED PRODUCT RETURN LINE EQUIVALENT ½ TUBING

GLAND RING BY-PASS FLUSH CONNECTION

FLOAT

TANK COVER

WATER LEVEL MUST BE ABOVE COILS

COILS

DURA THERMOCOOLER TANK

TANK DRAIN

HOT PRODUCT TO DURA THERMOCOOLER, EQUIVALENT ½ TUBING

STUFFING BOX

CLOSE FITTING THROAT RESTRICTION DEVICE

DURA CIRCULATING RING

SUPPLIED BY DMC

Figure 6-13. (a) Typical piping arrangement for dura thermocooler with dura circulating ring. (*Courtesy of Durametallic Corp.*) (b) typical piping arrangement for dura heat exchanger with circulating ring. (*Courtesy of Durametallic Corp.*)

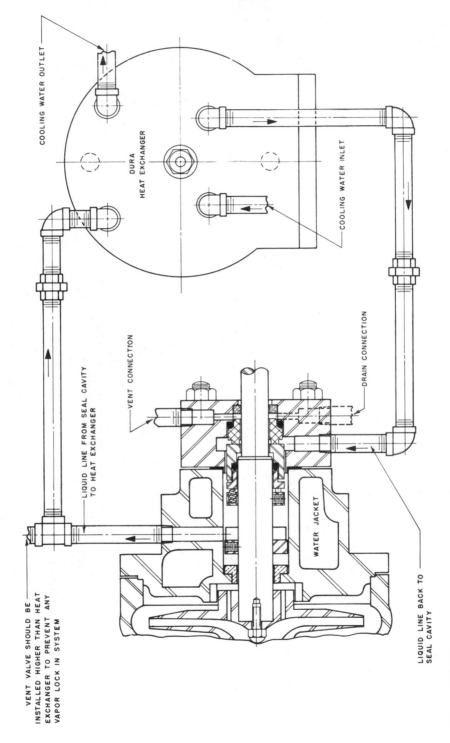

COOLING WATER OUTLET

DURA
HEAT EXCHANGER

COOLING WATER INLET

VENT VALVE SHOULD BE
INSTALLED HIGHER THAN HEAT
EXCHANGER TO PREVENT ANY
VAPOR LOCK IN SYSTEM

LIQUID LINE FROM SEAL CAVITY
TO HEAT EXCHANGER

VENT CONNECTION

DRAIN CONNECTION

WATER JACKET

LIQUID LINE BACK TO
SEAL CAVITY

Figure 6-13(b).

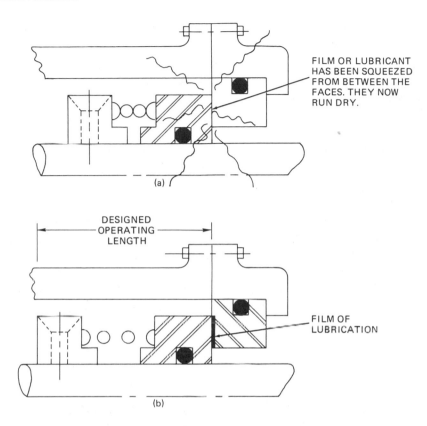

FILM OR LUBRICANT HAS BEEN SQUEEZED FROM BETWEEN THE FACES. THEY NOW RUN DRY.

(a)

DESIGNED OPERATING LENGTH

FILM OF LUBRICATION

(b)

Figure 6-14. (a) Too much compression; (b) seal installed at proper operating length (this ensures a desirable face loading). *(Courtesy of Sealol Inc.)*

To determine that the pump is ready for the seal, the concentricity of the shaft within the pump must be proven. This is verified with a dial indicator showing that the pump shaft has no more than 0.003 in. radial runout. The face of the seal ring must also be flat and its seat in the pump face must be checked again with the dial indicator to confirm such flatness. Checking the runout and squareness of the face and shaft are important, especially on a pump that has been in service, and where seal problems have developed.

The installation of the O ring, bellows, or V packing must be done carefully. Any sharp edges on the shaft, including any on the end of the shaft, a burr or rough spots on the shaft should be removed by rounding off the edges or removing shaft burrs with a crocus cloth. If you can feel any rough spots on the shaft, they will probably damage the O rings.

To position the seal for the correct operating length, you will need a reference point on the shaft. The operating length is the distance from the face of the seat ring to the back of the seal. When the seal is at the correct operating length, the springs will be partially compressed. The operating length is indicated on the drawing and instructions included with every mechanical seal.

The handling of the seal faces is critical to the proper operation of the seal. Do not allow them to get dirty, do not drop them, and do not scratch them. Treat the seal faces as you would a fine watch. It is important to remember that the seal faces have to be perfectly flat and polished.

The flatness of the seal face is determined by a testing instrument equipped with a special light and a clear glass precision optical flat. This device measures the flatness of the face in millionths of an inch and in wave lengths of light.

The failure of a mechanical seal must be investigated to determine the cause and to correct the defect. The reason or reasons for a seal failure are usually one of the following:

(1) incorrect installation
(2) defective parts
(3) an abrasive reaching the seal faces
(4) seal material not being compatible with the liquid being pumped
(5) wrong type of seal for the operation
(6) seal not protected against heat, abrasives or the product used
(7) failure of face lubrication
(8) pump not ready for seal (due to worn bearings, bent shaft, etc.).

The mechanical seal on a pump is an expensive precision product and removes pump maintenance from the hammer and chisel school of mechanics. Install the right seal in the correct manner and the results will be years of leak-free pump operation with the resultant savings that efficient operation of equipment will bring.

PUMPING PROBLEMS

Pumps are designed and built to meet the highest engineering standards; if properly selected, installed, and maintained, a pump can give years of trouble-free service. There are times when a pump fails to perform its designed function and inevitably there is a definite reason for such failure.

The following list of pump problems will be helpful in analyzing and correcting a particular problem.

Problem and Possible Cause

No liquid delivered

(1) Pump not primed.
(2) Speed too low: Check voltage and frequency.
(3) Air leak in suction.
(4) Discharge head too high.
(5) Suction lift too high.
(6) Impeller or discharge line plugged up.
(7) Wrong direction of rotation.
(8) Discharge valve closed.
(9) Plug in stuffing box or casing missing.

Not enough liquid delivered

(1) Air leaks in suction or stuffing box.
(2) Speed too low: Check voltage and frequency.
(3) Suction lift too high.
(4) Impeller or discharge line partially plugged.
(5) Not enough suction head for hot liquid.
(6) Total head too high.
(7) Pump defects:
    (a) Excessive ring clearances.
    (b) Damaged impeller or casing.
(8) Suction not submerged enough.

Not enough pressure

(1) Speed too low: Check voltage and frequency.
(2) Wrong rotation.
(3) Air or gas in liquid.
(4) Leaks in suction.
(5) Pump defects:
    (a) Excessive ring clearances.
    (b) Impeller diameter too small.

Pump works for a while, then loses suction

    (1) Leaky suction line.
    (2) Liquid seal plugged.
    (3) Suction lift too high.
    (4) Air or gas in liquid.
    (5) Air leaks in suction or at stuffing box.
    (6) End of suction line uncovered.

Motor runs hot (Note: Check actual temperature with thermometer.)

    (1) Pump taking too much power:
        (a) Speed too high.
        (b) Head lower than rating allowing pump to handle too much liquid.
        (c) Liquid heavier and more viscous than rating.
        (d) Pump defects:
           (i) Excess ring clearances.
          (ii) Stuffing boxes too tight.
        (iii) Rotor binding.
    (2) Electrical defects:
        (a) Voltage and frequency lower than rating.
        (b) Defects in motor.

Vibration

    (1) Starved suction.
        (a) Gas or vapor in the liquid.
        (b) Available NPSH not sufficient.
        (c) Inlet to suction line not sufficiently submerged.
        (d) Gas or vapor pockets in suction line.
    (2) Misalignment.
    (3) Worn or loose bearings.
    (4) Rotor out of balance.
    (5) Shaft bent.
    (6) Impeller plugged or damaged.
    (7) Foundation not rigid.
    (8) Impeller bore out of round.

Stuffing boxes overheat

    (1) Packing too tight.
    (2) Packing not sufficiently lubricated.

(3) Wrong grade of packing.

(4) Box not properly packed.

Bearings overheat

(1) Oil level too low.

(2) Improper or poor grade of oil.

(3) Dirt or water in bearings.

(4) Misalignment.

Bearings wear rapidly

(1) Misalignment.

(2) Shaft bent.

(3) Vibration.

(4) Lack of lubrication.

(5) Bearings improperly installed.

(6) Moisture in oil.

(7) Dirt in bearings.

OIL FUEL SYSTEMS

In all oil fuel systems a primary requirement is safe and adequate storage of the oil. Usually oil is stored in an iron tank placed either below ground or set on suitable supports above ground. The most economical type of storage tank is a buried one, which does not need the insulation required for an exposed tank, and it does not need the diking that is required around an above-ground tank. Governmental regulations require that, around any storage tank that poses potential pollution for surface waters, the tank must be surrounded with a dike that has sufficient capacity to contain any tank spillage.

Figure 6-15 illustrates a typical buried storage tank.

Provision must also be made for filling, pumping out and venting the tank, and heating the oil.

If the tank contains any grade of oil within the range of 4 to 6, the oil must be heated. The reasons for this are related to the viscosity of the oil (see Fig. 6-16):

(1) Heating helps pump the oil by reducing its viscosity, allowing the oil to flow more easily.

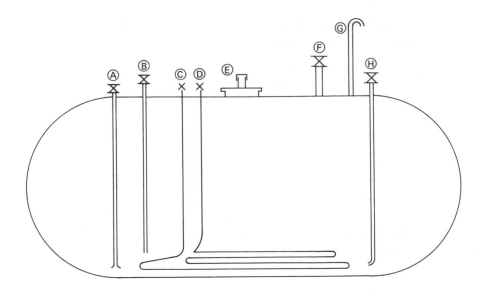

LEGEND
A   LOW SUCTION
B   HIGH SUCTION
C   STEAM HEATING INLET
D   CONDENSATE OUT
E   SAMPLING INLET
F   FUEL OIL RETURN
G   VENT
H   FILLING LINE

Figure 6-15. Fuel oil storage tank.

(2) Reduction in viscosity assists in the separation of any water
     or particulate matter in the oil, allowing it to settle to the
     bottom of the storage tank.

The assumption that if a little heating is good, a lot of heating is
better, is not applicable here. Always be aware of the temperature of
the oil in the tank. As previously noted, the flash point of the oil is
that point on the temperature scale at which the oil will give off volatile
vapors. If therefore, the temperature in the tank exceeds the flash point,
volatile vapors will be given off by the oil in the tank. These volatile
vapors contain a portion of the Btu value of the oil and so the Btu
value will be reduced in proportion to the release of volatile elements.
The settling of the water and sludge at the bottom of the tank, induced
by heating, requires inclusion of a high and low suction pipe in the
tank. All boilers should be operated at all times on the low suction
line. This practice insures continuity of operation by enabling a shift

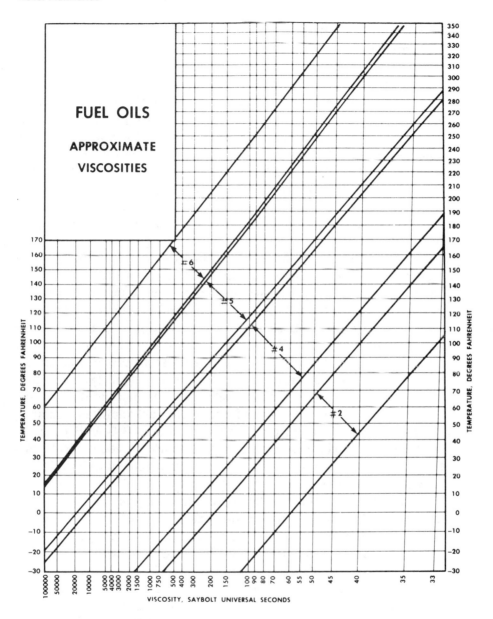

Figure 6-16. Fuel oil viscosity. *(Courtesy of Delaval Turbine Inc.)*

to the high suction line if the tank develops a leak from ground water or a delivery of fuel contains a large proportion of water. The difference in level between the high and low suction lines is about 12 inches. This allows for a rise in water level of 12 inches in the tank after you have shifted to the high suction line and allows time for corrective action, such as pumping out the water in the tank or obtaining another source of fuel.

It is important to be aware of the conditions at the suction line and, for this purpose, a sampling bomb must be used periodically. The sampling bomb is a device for retrieving a sample of the oil from the bottom of the tank. The unit is lowered into the tank by a chain, and when it touches bottom, the valve is opened, allowing the contents in the immediate vicinity to flow into the container. When the bomb is withdrawn the valve closes. It is usual to find a mixture of oil, sludge, and water in the sample. On occasion the sample will contain 99% water, indicating that action must be taken to reduce the quantity of water.

It will be noted that the line for filling the tank is at the opposite end of the tank from the suction line and also that the return line to the tank returns the oil to the surface of the liquid level (see Fig. 6-17). These pipes are so positioned in order to circulate the oil in the tank. The filling line delivers the oil to the bottom of the tank and an elbow at the bottom directs the incoming oil along the tank bottom, tending to scour it. After the oil is drawn from the bottom of the tank, any excess pumped is returned to the top of the tank and thus circulates. This circulating action helps in the reduction of water and sludge and, with the assistance of one of the fuel additives available, the accumulation of bottom sludge and water (BS & W) can be reduced. Pumping a tank out to clean it is not an economical practice; you lose too much oil. With careful management the need to do so will be eliminated.

The purpose of the vent line is to allow for the displacement of air when filling the tank and to prevent the formation of a vacuum when the tank is pumped out. With a return elbow at the top of the vent for weather protection the outlet is covered with a fine-mesh copper screen, which acts as a spark arrester.

From the suction pipe in the tank a line leads to the boiler room and a first set of strainers. The strainers, a duplex set, are fitted with coarse screens and are designed to remove any large particles in the oil. Compound gauges, which show pressure and vacuum, are fitted on either side of the strainers ahead of the pump. These gauges will usually indicate a vacuum reading and, when the strainer is cleaned and first put in service, the gauges should show the same reading. By observing this reading, the operator can be aware of the conditions prevailing with a clean strainer. As the strainers become dirty, a differential will show between the gauges; the gauge nearer the pump

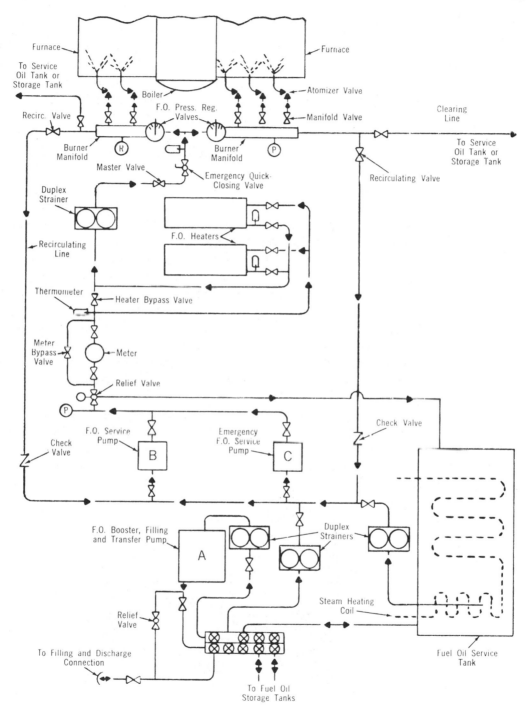

Figure 6-17. Fuel oil system. *(Courtesy of Delaval Turbine Inc.)*

or upstream from the strainer will show a reduction in vacuum. This indicates that the strainer is becoming dirty and needs cleaning. A positive indicator of the need to clean can save the operator the time and trouble of cleaning when not necessary, and the use of the gauges can indicate trouble prior to a reduction in oil pressure and a failure of the boiler plant because of dirty strainers. Also at the strainers is a thermometer which shows the temperature of the oil and also provides a check on the temperature in the tank. The heat of the oil in the tank must be controlled to prevent its vaporizing when subjected to the vacuum created by the pump.

From the strainers, the oil enters the fuel pump, which is always a positive displacement pump, and is pumped into the fuel oil heater (see Fig. 6-18). On the discharge side of the pump is a safety valve to relieve any excess pressure caused by a line blockage or a shutting down of the burner. Any discharge from the safety valve is directed either to the suction line of the pump or returned to the fuel tank.

The fuel heaters, of shell and tube design, are usually steam heated. On the discharge of the heaters is a thermometer and thermowell. The thermometer gives a visual reading of oil temperature and the thermowell contains a bulb that is connected to the capillary tubing, which in turn controls the flow of steam into the heater via a steam control valve.

The discharge from the heater passes into a second set of strainers, called the fine strainers, which are fitted with a fine screen and will remove the finer particulates that have passed through the first strainer. These strainers are also equipped with gauges, showing pressure only,

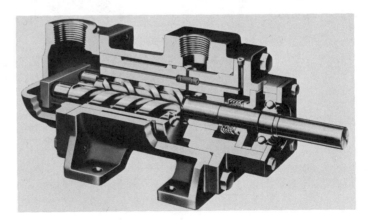

Figure 6-18. Fuel oil pump. *(Courtesy of Delaval Co.)*

which serve the same purpose as the compound gauges. A differential reading indicates the need to clean the strainers.

On the discharge line of the strainers ahead of the burner, it is good practice to provide an electric heater. This unit, usually a resistance element enclosed in a shell, is used to heat the oil when it is necessary to start the boilers from cold and no steam is available for heating. This unit should be equipped with a thermostat to regulate the heat and when not needed should be shut off from the power source. Carbonization of the oil on the elements frequently makes them useless when they are needed most. To alleviate this condition, install this heater as a bypass around the main line, and only allow oil to flow through the unit when needed.

The oil line is continued to the boiler and into the fuel oil control at the burner.

THE REDUCING VALVE

In all power plants, the ability to use steam, air, water, and fuel oil at varying pressures throughout the system is desirable economically. The generation of steam at a high pressure and subsequent reduction for other uses makes a pressure-reducing valve an important auxiliary in the power plant.

All pressure-reducing valves operate on what is known as the "constant heat expansion" principle, wherein expansion of the vapor or fluid is accomplished by throttling or restricting the flow through an orifice from a high to a lower pressure. The only work being done is the overcoming of friction through the orifice. The heat content would be the same at a high pressure as at a lower pressure.

Because of the same heat content at the lower pressure when steam pressures are reduced, the reduced pressure will be somewhat superheated and have a temperature higher than the corresponding steam pressure. Experiments have shown the following results:

Gauge Pressure (psi)	Temperature (°F)	Reduced Pressure (psi)	Temperature	Degrees of Superheat (°F)
160	370	100	349°	12°
200	384	120	369°	20°
200	384	10	341°	128°

To achieve the maximum effectiveness of a reducing valve, it is important that the valve be properly sized for its use. An oversized valve will cause an excess of *wire drawing* or cutting of grooves, between the valve and seat, shortening the life of the valve, and a valve sized too small will cause a loss of pressure on the downstream of the valve at high flow rates.

All manufacturers provide data on sizes and flow rates for their particular product, but all sizing of valves is based on the basic relationship that the sum of the velocity and static pressures of the fluid entering the valve must equal the sum of the velocity and static pressures exiting from the valve plus friction losses. Therefore, flow rate is proportional to the square root of the pressure drop; the basic equation being:

$$q = C_v \frac{\sqrt{(P_1 - P_2)}}{G}$$

q is gallons flowing per minute, P_1 and P_2 are pounds of pressure per square inch and G is the specific gravity of the fluid. The constant C_v is the valve sizing coefficient established for each valve by tests, and is defined as the number of gallons of water that will pass through a given *flow restriction* with a pressure drop of one pound per square inch—for example, a valve with a C_v of 25 gallons of water to flow each minute, resulting in a pressure drop of one pound.

The relationship between the pressure drop and the velocity of fluid flowing through any valve is illustrated in Fig. 6-19. First velocity increases, then it decreases as it moves through the valve. The increased velocity requires a lower pressure. The point of maximum fluid *jet contraction* is called the "vena contracta," and beyond this point the flow returns to its former area valve in the recovery region. It is in this region that most friction and turbulence losses occur, which accounts for most of the pressure losses between inlet and outlet pressures.

In addition to C_v in sizing, it is important to remember that the flow rate through a valve increases as the pressure drop increases, up to a certain point. Beyond this point, no increased flow takes place, regardless of downstream pressure.

A phenomonon called choking occurs when the valve reaches "saturation pressure," which is the pressure at which a liquid will boil at a given temeperature. Since the lowest pressure occurs at the vena contracta, if this pressure corresponds to the liquid saturation pressure, vaporization will occur. Vaporization of the liquid produces choking.

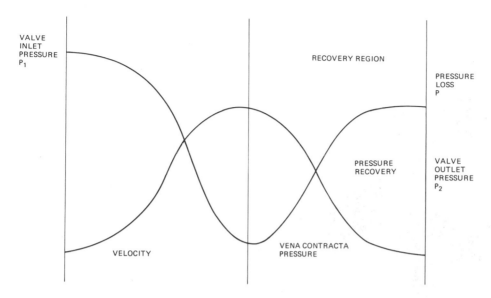

Figure 6-19. Valve velocity and pressure relationships.

Experiments confirm that the principle of choking also applies to gases. As the pressure drops accelerates, flow becomes greater, but the rate of increase diminishes at higher pressure drops until no additional flow will pass through the valve.

When hot water at or near saturation temperature flows through a reducing valve, a mixture of water and steam is created at the outlet of the valve. Examples of such valves are shown in Figs. 6-20, 6-21, and 6-22.

Any use of a reducing valve requires full knowledge of the flow rates, pressure differentials, and temperatures involved in an installation. The problems involved in valve sizing have been reduced by the use of the flow coefficient C_v, which is based on the fundamental formula of fluid flow $V = \sqrt{2gh}$ and is not restricted to any particular valve.

FEEDWATER HEATER

The heating of the boiler feedwater before pumping it into the boiler is necessary for the following reasons:

(1) Cold water fed to the boiler will require additional fuel to raise its temperature.

Figure 6-20. Preset reducing valve. *(Courtesy of Jordan Valve Co.)*

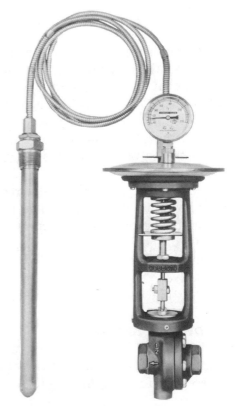

Figure   6-21. Temperature-regulating valve. *(Courtesy of Jordan Valve Co.)*

Figure 6-22. Diaphragm-operated pressure-reducing valve. *(Courtesy of Jordan Valve Co.)*

(2) Cold water carries more impurities into the boiler than hot water, and therefore increases the probability of internal scale formation.

(3) Cold water fed to a boiler produces a localized cold spot, causing unequal expansion of the boiler shell with the possibility of straining joints and tubes, and generating leaks.

(4) If the water is heated before entering the boiler, the effective capacity of the boiler to produce steam is increased.

The method of heating feedwater can be by a closed or surface heater or by an open deaerating heater.

The closed heater uses high-pressure exhaust or live steam to provide heat to a shell-and-tube heat exchanger. The temperature to which the feedwater can be heated by this method is only limited by the pressure of the steam supplied, whether live or exhaust steam.

An open deaerating feed heater is used in the majority of boiler plants and serves a two-fold purpose. It provides heat to the feedwater and also removes the noncondensable gases, most particularly oxygen and carbon dioxide. This removal of oxygen and other gases aids in the protection of the boiler against corrosion.

A typical deaerating feedwater heater has a reservoir tank to provide a supply of feedwater and is situated at the highest point in the boiler room. The height is necessary to provide a positive suction head to the feedwater pump.

A dome or head on the reservoir tank contains trays over which the incoming condensate returns or additional cold makeup water cascades. Live or exhaust steam rising through the dome "scrubs" the water as it descends in thin sheets and carries with it the noncondensable gases to the atmosphere through a vent pipe.

Savings in water consumption can be achieved if a vent condenser is used on the atmospheric vent. This condenser, of a shell-and-tube type, allows the mixture of gases and steam to flow through the tubes and the incoming makeup feedwater flows around the tubes, effectively condensing any steam escaping to the atmosphere and returning it to the reservoir tank.

The deaerating heater has float and level controls for the reservoir tank and thermometers and pressure gauges to indicate the temperatures and pressures involved. The usual temperature of the feedwater from an open deaerating heater is 220°F − 230°F with a corresponding tank pressure of 4 to 5 psig.

In order to maintain temperature and the deaerating function when the returning condensate is at a low temperature, a live steam connection is made to the deaerating dome. This live steam is controlled by a regulating valve so designed that if the pressure in the reservoir tank falls below a set pressure, live steam is admitted to the tank to deaerate and heat.

When annual maintenance is performed, it is important to clean out any debris or scale that has been precipitated in the tank. The alignment of the cascade trays or spray nozzles should be checked to make sure that the water entering the deaerator is in a stream that fills the dome, ensuring maximum scrubbing of all feedwater.

EXERCISES

1. Why does steam become superheated after passing through a reducing valve?
2. What is constant heat expansion?

3. A boiler room feed pump has a static head of 25 ft. What is the pressure at the pump suction?

Ans.: 10.8 lb/in²

4. A pump is situated 5 ft above the liquid level of a tank. If the pump discharges into a tank 10 ft above the suction of the pump, what is the total dynamic head?

Ans.: 15 ft

5. What is cavitation? A pump discharges the equivalent of 60 hp. If the pump is driven by a 75 hp motor, what is the efficiency of the pump?

Ans.: 80%

6. A weight is dropped from a height of 200 ft. With what velocity will it hit the ground in feet per second?

Ans.: 113.373 ft/s

7. A centrifugal pump has an impeller 9 in. in diameter. The total head is 88 ft. At what r/min must the pump be driven to overcome the head?

Ans.: 1761.19

8. A pump running at 3600 r/min delivers water at 160 g/min. If the rotational speed of the pump is cut by 25%, how many g/min will the pump deliver?

Ans.: 120 g/min

9. A pump shaft is 3 in. in diameter. If the stuffing box is 4 in. in diameter, what size of packing is needed?

Ans.: ½ in.

10. What is a lantern ring? How is it used?

11. What are the advantages of a mechanical seal?

12. How can a mechanical seal fail? Be specific.

13. Why is oil heated when in storage?

14. What is the flash point and the fire point of fuel oil?

15. Why would you use a sampling bomb?

16. In your own words, describe a fuel oil system to burn No. 6 oil.

17. If the temperature of the feed water is too high, what happens to the pump capacity?

18. Why is boiler feed water deaerated?

MULTIPLE CHOICE QUESTIONS

1. Atmospheric pressure at sea level is:
   a. Gauge plus atmospheric
   b. 14.7 lb/in²
   c. 15 psi

2. Theoretical suction lift for a pump at sea level is:
   a. 3.3957
   b. 33.3957
   c. 333.957

3. When a liquid boils at a pressure below atmospheric, the liquid has reached its:
   a. Boiling point
   b. Pressure point
   c. Vapor pressure

4. A pump that requires an internal or external relief is a:
   a. Centrifugal pump
   b. Gear pump
   c. Steam pump

5. When a centrifugal pump is throttled at the discharge:
   a. The amperage will fall
   b. The motor will overheat
   c. Nothing happens

6. When pumping oil for burner service, always use:
   a. High suction
   b. Low suction
   c. A centrifugal pump

7. A spark arrestor is always needed on a:
   a. Fuel line
   b. Suction line
   c. Vent line

8. Number 6 fuel oil requires _____ to reduce its _____.
   a. straining, specific gravity
   b. heating, flash point
   c. heating, viscosity

9. Reducing valves operate on:
    a. The constant heat expansion principle
    b. Pressure–volume relationship
    c. Desuperheating factor

10. The lowest pressure through a valve occurs at the:
    a. Vena contracta
    b. Vona contracta
    c. Vana contracta

7

Steam Trapping

There are two classes of heat in steam. One is sensible heat and the other is the latent heat of vaporization. To make full use of all the Btu's contained in steam when it is being utilized to heat a vessel, jacketed tank, or for space heating of any kind, it is necessary to allow the steam to condense when it is in contact with the object being heated. To provide the necessary residence time, the flow of steam away from the vessel must be restricted. From the earliest days of its use, steam has been "trapped" either by means of an orifice or a partially closed valve.

The orifice trap still in use today consists essentially of a variable orifice in the condensate line. This form of trap works well, but has the objectionable feature of being open continuously and therefore perpetually wasting energy. To overcome this drawback various types of steam trapping devices have been developed. All work well for their intended applications, and provide the orifice that has always been used, but the modern device can be characterized as an orifice with a "brain."

The types of traps available and in general use today are the inverted bucket, float thermostatic, thermostatic, and disk traps.

The early use of condensate drainage was accomplished with an orifice-type trap and a ball float trap, whereby a container gradually filled with water, a float rose with the level of the water and opened a valve, which allowed the water to be discharged. A problem with this type of trap was that it would become air bound, holding the trap in the closed position. This air was released by opening a pet cock, allowing the trap to function properly.

Many companies today make steam traps of all shapes and sizes. One of the leading companies is the Armstrong Machine Works of Three Rivers, Michigan, which has manufactured steam traps for 75 years. Many of the advanced techniques used in steam trapping are the result of research and development at Three Rivers.

The most widely used and adaptable steam trap today is the inverted bucket trap (see Fig. 7-1). This trap was invented in 1910 by a friend of Adam Armstrong named Otto Arner. Adam Armstrong started his business career making bicycle spokes and potato diggers. With the float and open top bucket traps then being used in his factory, the air in the system could not be released. Otto Arner suggested inverting the bucket and drilling a hole in its top to release the air with the condensate. On October 17, 1911 a patent was issued on an inverted bucket trap using leverage to multiply the power of the bucket.

Figure 7.1. Section through an inverted bucket trap. Note air vent in top of bucket. *(Courtesy of Armstrong Machine Works)*

Figure 7-2 shows the original Armstrong patent of 1911 compared with a modern trap. In Figure 7-3, the trap is installed between a steam unit and the condensate return header. The bucket is down and the valve is wide open. The initial flood of condensate enters the trap and flows under the bottom edge of the bucket to fill the trap body and completely submerge the bucket. Excess condensate discharges through the wide open valve to a condensate header.

When steam reaches the trap it collects at the top of the bucket, imparting buoyancy. The bucket then rises and lifts the valve towards its seat. The flow of condensate snaps the valve tightly shut. Air and carbon dioxide gas pass through the small vent hole at the top of the bucket and collect at the top of the trap. Any steam passing through the vent is condensed by radiation.

When the entering condensate brings the condensate level slightly above the neutral line, the bucket exerts a slight pull on the lever. But the valve does not open until the condensate level rises to the opening line for the existing pressure differential between the steam and the condensate return header.

When sufficient condensate has collected in the trap, the weight of the bucket multiplied by the leverage exceeds the pressure holding the valve on its seat. The bucket then sinks and opens the outlet

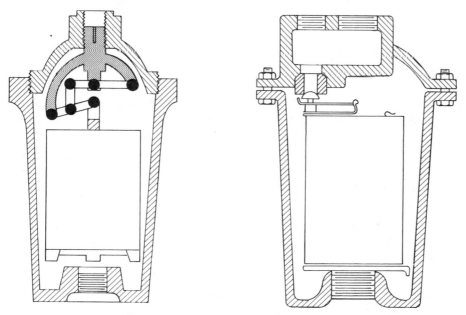

Figure 7-2. 1911 original vs. modern trap.

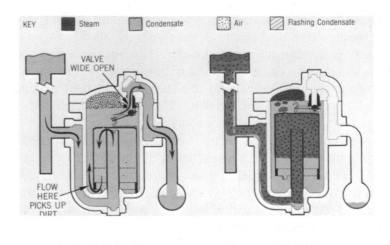

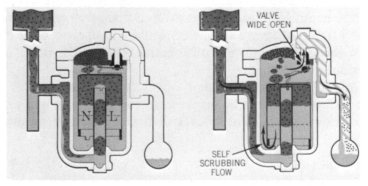

Figure 7-3. Action of inverted bucket trap. *(Courtesy of Armstrong Machine Works)*

valve. Accumulated air is first discharged, followed by the condensate. Discharge continues until steam enters the trap to float the bucket, and then the cycle is repeated.

While the inverted bucket trap has the capability of serving the vast majority of trapping needs, there are operating conditions where other types of traps are needed. The float thermostatic trap is useful where the high-capacity venting of air and CO_2 are required along with a continuous discharge of condensate. The float thermostatic trap (see cross section in Fig. 7-4) embodies a float-actuated valve and a balanced-pressure thermostatic air vent. When condensate accompanied by large volumes of air and CO_2 enter the trap, the float holds the valve in the closed position. Any air brought into the trap is immediately vented through the thermostatic air vent. As the temperature of the entering condensate rises, the piping system is

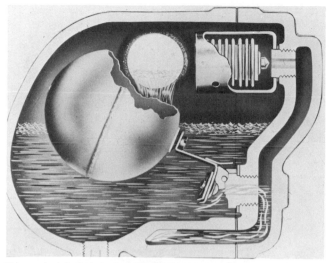

Figure 7-4. Section through a float-thermostatic steam trap. *(Courtesy of Armstrong Machine Works)*

relieved of cold condensate and live steam pushes any trapped air out of the system; the thermostatic air vent starts to close and will remain closed when steam at any pressure within the operating range of the trap is present.

The thermostatic bellows trap operates on the same principle as the bellows in the float thermostatic trap. The bellows trap (Fig. 7-5) works as follows: (1) on startup, the condensate and air are pushed ahead of the live steam directly through the trap. The thermostatic bellows element is fully contracted and the valve remains wide open until steam enters the trap; (2) as the temperature increases it quickly heats the bellows element, increasing the vapor pressure inside. When this pressure becomes balanced with the system pressure in the trap body, the spring effect of the bellows causes the element to expand, closing the valve.

When the temperature in the trap drops a few degrees below the saturated steam temperature, the bellows contract, opening the valve to allow condensate flow.

The disk trap has become popular because of its light weight, relatively low price, and its ability never to fail in the closed position. The disk trap functions as follows: The condensate and air entering the trap pass through the heating chamber around the control chamber and through the inlet orifice. This flow lifts the disk off its seat, and the condensate flows through to the outlet passages. When steam reaches the disk, increased flow velocity across the face of the disk

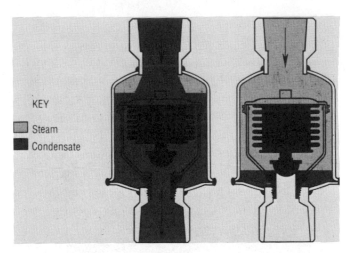

KEY

Steam

Condensate

Figure 7-5. Thermostatic bellows trap. *(Courtesy of Armstrong Machine Works)*

reduces pressure at this point and increases pressure in the control chamber, and the disk closes the orifice. Controlled bleeding of steam from the control chamber causes the trap to open. If the condensate is present it will be discharged. The trap recloses in the presence of steam and then continues to cycle. The disk trap is useful in a power plant where small size, ease of selection for various pressures, and ability to operate on light loads are primary requirements. Its main fault lies in its short life due to rapid wear of the disk, causing excessive steam loss and therefore energy waste.

Steam trapping is used to extract maximum heat value from steam being utilized in a manufacturing process, but for the power plant operator it provides the valuable function of keeping the steam mains and headers free of condensate.

It is usual when first starting up a steam system or starting it up after a prolonged shutdown to have what is known as a supervised warmup, where drip leg valves are opened to the atmosphere and allowed to flow freely for rapid removal of condensate. The valves are closed when all mains are up to working pressure. It is, however, possible to raise steam on a boiler and then with all main stops and header valves open, allow the steam to gradually work its way through the system—all drip legs and traps functioning to remove the condensed steam.

The number of traps necessary in any system is determined by the complexity of the system and the number of dead ends or raises or other natural drainage points.

At all these points, drip legs should be provided in order to let condensate escape by gravity from a high-velocity steam flow and to store the condensate until the pressure differential is great enough for the trap to discharge it.

A functioning trap is a necessity on all steam mains and headers in order to assure "dry" steam and reduce the possibility of water hammer in the steam piping system.

Water hammer is not really water hammering in the system but results from water or condensed steam coming in contact with live steam. If, for instance, a steam valve is opened into a relatively cold line, the steam will immediately condense and form water in the steam main. As steam continues to flow into the main, the water already present condenses the steam as soon as the two elements, steam and water, come into contact. As the steam condenses a vacuum is created, which is immediately filled by the live steam under pressure. This sudden rush of steam causes the rapping or hammering familiar to all operators. A functioning drip trap combined with slow opening of the steam valve will do much to eliminate water hammer.

All traps, regardless of size or functions, have to be sized for the job they are to do, and the most important factors in selecting a steam trap are the pressure differential under which it will work and the amount of condensate it will be expected to handle.

The pressure differential around any steam trap is the difference between the inlet and outlet pressure.

The inlet pressure can be the boiler or steam-main pressure, turbine back pressure or the pressure in a main reduced from boiler pressure. Outlet pressure can vary, depending upon the point of discharge. Such pressure falls into the following categories: atmospheric pressure (such as a tracer line trap discharging into the atmosphere); below atmospheric pressure (as into a return line on a heating system with a condensate vacuum pump or into the condenser on the low-pressure end of a turbine); above atmospheric pressure (due to pipe friction losses); controlled back pressure (as into a feed heater); elevating condensate (to a return header) or siphon drainage (when the steam trap is above the point being drained). All these factors have to be considered when determining the size of a steam trap.

To calculate the amount of condensate to be handled for a drip trap it is necessary to know whether, when raising steam, you will have a supervised warmup or an automatic warmup, the pipe sizes involved, and the initial and final temperature of the pipes. It is important to note that the amount of condensate produced during the warmup period will exceed the amount of condensate produced when at operating temperature, especially in insulated piping.

Most manufacturers of steam traps and trapping devices provide all users and potential customers with practical data on a full range of traps in order to make the selection relatively easy. However, if no such data is available, the following formula can be used to size a steam trap for a drip leg on a piping system:

$$C = \frac{W\,(t_1 - t_2)\,0.114}{L}$$

where: C = Amount of condensate
W = Total weight of the pipes
t_1 = Final temperature
t_2 = Initial temperature
L = Latent heat of steam at final temperature
0.114 = Specific heat of steel pipe

When the temperature of the pipe exceeds 212°F, the amount of condensate being generated will decline. Therefore, calculate the pounds of condensate generated until the pipe reaches 219°F, the equivalent of 2 lb/in². Divide the figure by the number of minutes it will take for the pipe to reach 2 psi or 219°F and multiply by 60 to calculate pounds of condensate per hour. Select a trap to handle that amount of condensate, allowing a safety factor of 2 to 1 for lines connected directly to the boiler and carrying full boiler pressure and a factor of 3 to 1 for drip traps located ahead of reducing valves and shutoff valves.

Regardless of the type of trap, there are certain basic rules that apply to all trap installations.

1. Use strainers ahead of all traps.
2. Install trap so that it is accessible for checking and repair.
3. Install below drip point whenever possible.
4. Install close to the drip point.
5. Use shutoff valve ahead and downstream of trap (see Fig. 7-6).
6. Do not use bypasses around traps except for process equipment where continuous operation is essential.

7. If only one union is used it should be on the discharge side of the traps.
8. Use check valve on discharge of trap when condensate has to be elevated.

Once a steam trap is installed, it is essential that a check be made on its operation at reasonable intervals to be sure that you are discharging any condensate and that the trap is not damaged or worn, either of which causes a continuous blowing of live steam into the condensate return system with a resultant waste of energy. The test valve (Fig. 7-7), is best but not always practical; but if used, the return line valve should be closed and test valve opened. An intermittent discharge indicates the trap is working correctly, a continuous steam blow indicates a defective trap.

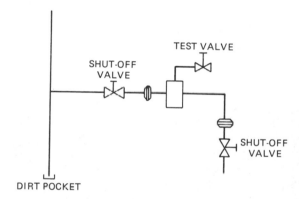

Figure 7-6. Typical inverted bucked trap hook-up with test valve.

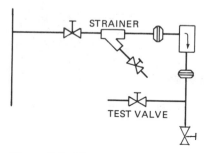

Figure 7-7. Thermostatic trap with test valve.

Checking the temperature on either side of the trap is a good indication of whether a trap is working correctly. Using a surface pyrometer, take a reading on either side of the trap. A difference in temperature should be noted between inlet and discharge. If there is high back pressure in the discharge of the trap, the trap could be working correctly, but the high back pressure would cause little temperature differential.

The best method of checking any trap is with an ultrasonic listening device (Fig. 7-8). A method for listening to a trap work via a steel rod was referred to earlier, but with advanced technology it is possible to actually hear and observe a steam trap working. One such ultrasonic device, (as pictured) is a compact hand-held instrument appropriately called a "Trouble Shooter." A little practice will enable the boiler operator to detect the sound of a steam trap at work and to check the traps at regular intervals without taking any unit out of service.

It is a vital part of the engineer's daily work to insure that all equipment under his control works efficiently. Steam traps are vital parts of that equipment.

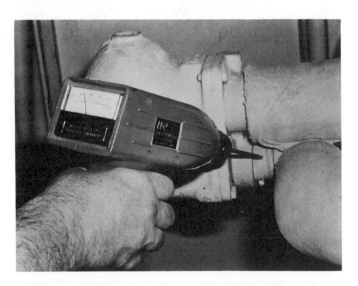

Figure 7-8. Using the trouble shooter. *(Courtesy of U.E. Systems)*

EXERCISES

1. What is a steam trap?
2. Describe the action of an inverted bucket trap.
3. In what type of steam system is the float thermostatic trap most effective?
4. What can cause a bellows trap not to work?
5. Why does a disk trap waste energy?
6. What is a supervised warmup? How does it differ from any other type of warmup?
7. What is water hammer?
8. What is pressure differential around a steam trap?
9. Describe how you would check to find out if a steam trap is working correctly.

MULTIPLE CHOICE QUESTIONS

1. Steam traps are used to ensure full use of the:
   a. Specific heat
   b. Sensible heat
   c. Latent heat

2. Float thermostatic traps are used where large quantities of _____ must be discharged:
   a. water
   b. air
   c. gas

3. A thermostatic bellows trap operates on the same principle as the bellows in a:
   a. Disk trap
   b. Float trap
   c. Bucket trap

4. The disc trap is popular because of:
   a. Light weight
   b. Low cost
   c. Both a and b

5. A steam system can be started up after a prolonged shutdown using a:
   a. High fire-up
   b. Supervised warm-up
   c. Low fire system

6. Good drain traps help to reduce incidents of:
   a. Carryover
   b. Low water
   c. Water hammer

7. A factor in sizing a steam trap is:
   a. Pressure differential
   b. Material of construction
   c. Type of trap

8. One of the basic rules when installing a trap is:
   a. Use valves
   b. Use strainers
   c. Use bypasses

9. Continuous blowing of steam through a trap:
   a. Is necessary
   b. Wastes energy
   c. Wastes water

10. A trap can be tested with:
    a. Ultraviolet
    b. Ultrasonic
    c. Ultramarine

8

Water Treatment

In the operation of a steam power plant, water quality is a critical factor. Pure water, in its natural state, is odorless, tasteless, and colorless and is the most universal of solvents. Water's capability as a solvent makes its treatment and analysis integral functions of boiler operation.

The various types can be classified as surface water (from streams, rivers, ponds, lakes, and reservoirs) and ground water (from springs and shallow and deep wells).

Water is evaporated from the earth's surface by the sun and returns to earth in the form of rain, hail, and snow. In its descent through the atmosphere it absorbs oxygen, nitrogen, and carbon dioxide. On the ground it washes silt and sand into lakes and rivers. Organic matter from plants and trees is absorbed by the water and accumulates in reservoirs and lakes.

The type that percolates through the soil into ground water deposits dissolves minerals and chemicals from the soil. Impurities that water picks up can be divided into two classes: suspended solids and dissolved solids. Suspended solids are those that do not dissolve in water and can be removed by filtration. Dissolved solids are those that break up in water and cannot be removed by filtration. Gases may dissolve in water, but unless they combine chemically with other impurities in solution, they can be expelled by boiling the water.

The chemicals and minerals that create problems in the operation of a boiler system are as follows:

Name	Chemical Formula	Difficulties Caused
Carbon dioxide	CO_2	Corrosion in steam and condensate lines.
Sulfate	(SO_4)	Adds to solids content of water, combines with calcium sulfate scale.
Chloride	CL	Adds to solids content, increases corrosive character of water.
Nitrate	(NO_3)	Adds to solids content.
Silica	$Si\,O_2$	Source of scale in boilers and cooling systems and deposits on turbine blades.
Oxygen	O_2	Corrosion of water, condensate lines, heat exchanger and boilers.
Calcium	Ca	Source of scale in boilers.
Magnesium	Mg	Source of scale in boilers.

The formation of scale on boiler heating surfaces is the most serious problem in steam generation. The object of the processes for treating water before it enters the boiler is to reduce the amount of scale and sludge-forming deposits. However, since no external treatment, regardless of efficiency, can remove all harmful chemicals, we must provide internal treatment to boiler water.

The primary cause of scale formation is a decrease in solubility of the salts accompanying an increase in temperature. Consequently, the higher the temperature and pressure in a boiler, the more insoluble are the scale-forming salts.

Deposits of scale occur in two basic ways. First, water enters the boiler and, being raised in temperature, deposits calcium carbonate (as $CaCO_3$) on the boiler steel which bakes in place as scale. Secondly, scale is also deposited from local supersaturation of the boiler water, especially in boilers with sluggish circulation. The thin film of boiler water immediately adjacent to a heating surface tends to become more concentrated than the main body of water in the boiler. The solubility of the salts in this area is reduced and deposits scale on the heating surface.

Scale in a boiler creates a problem because of the low degree of heat conductivity possessed by all forms of scale. The presence of the scale provides an insulating barrier to the flow of heat and reduces boiler efficiency. Stack temperatures will increase because the boiler absorbs less heat from the products of combustion.

Studies have shown large losses in efficiency of operation because of scale deposits on the waterside of a boiler. Even a 2% to 3% reduction in heat transfer results in a substantial monetary loss because of increased fuel costs.

But more important is a possible overheating of the boiler metal due to the lack of heat transfer caused by scale deposits. The resultant damage caused by overheating can result in burned-out tubes and blisters on steam and water drums.

The salts of calcium and magnesium are the most common source of boiler scale. The use of external treatment facilities such as demineralization, brine softening, and lime-soda or hot phosphate softening are designed to remove calcium and magnesium salts from boiler water. Internal treatment seeks to prevent deposits of residual salts and to maintain clean boiler surfaces.

The precipitation of calcium carbonate to form boiler scale readily takes place when the boiler feedwater contains any appreciable amount of calcium bicarbonate. The presence of calcium sulfate, a more soluble salt than calcium carbonate, in boiler scale indicates inadequate treatment. All methods of internal chemical treatment rely on precipitating the calcium salts into some form other than calcium sulfate.

Magnesium salts present in feedwater are usually more easily prevented from forming scale than the calcium salts. Normally magnesium is precipitated in the form of magnesium hydroxide. The magnesium salts found in boiler scale are magnesium hydroxide, magnesium silicate, and magnesium phosphate. Magnesium phosphate is not a very hard scale, but magnesium silicate forms an extremely hard scale, usually on places of high heat transfer.

Both calcium and magnesium cause problems in boiler operation that may be compounded by deposits of silica, which may combine with them to form complex deposits, especially in boilers operating at high pressure. The prevention of boiler scale due to silica can usually be solved by external treatment, but the maintenance of high concentrations of alkalinity and high residual phosphates in the boiler can aid in prevention of silica scale.

For the internal treatment of boilers and boiler water, many unusual and varied materials and methods have been used. When marine boilers were fed with sea water it was to be expected that scale would form. After cleaning, the internals on the water side would be coated with kerosene to make the scale easier to remove. Various types of coatings were applied to prevent scale formation and make the removal of scale easier.

For modern industrial plants, the trial-and-error methods previously used are out of date. Specific treatments have been developed for particular problems. To prevent calcium and magnesium salts from baking on boiler heating surfaces, internal treatment involves precipitating the salts into a harmless sludge and keeping this sludge in a fluid form to be removed via the boiler blowdown. Calcium is more difficult to handle than magnesium because magnesium is readily precipitated by the alkalinity of boiler water as sludge in the form of magnesium hydroxide.

The most common chemicals used to precipitate soluble calcium salts are the sodium phosphates, trisodium phosphate, and disodium phosphate. Once these chemicals enter the boiler they react with calcium to form tricalcium phosphate, a flocculent precipitate. To insure that this action takes place, a high alkalinity must be maintained in the boiler. At a pH of 9.5 or less the calcium will not be precipitated properly.

Where it is desired, in some waters with very high alkalinity, acid phosphates such as sodium metaphosphate and monosodium phosphate can be used to reduce the pH. The action they cause when entering the boiler is exactly the same as the trisodium phosphate, but they will reduce the alkalinity.

Once the calcium and magnesium salts have been precipitated, it is necessary to keep them in a fluid form in order to blow them out of the boiler. This is accomplished with the use of organic compounds such as the tannins, lignins, starches, and seaweed derivatives. Tannins, extracted from wood and bark, have a high molecular weight and are chemically complex. Lignins are also complex organic substances. With starches and seaweed derivatives they form the basis for the polymer chemistry that allows total solids concentrations to be dissolved in boiler water without scale formation. All of these organic agents form a coating around the inorganic precipitates, decreasing their tendency to adhere to the boiler surface. The organic agents are also important in effecting crystal formations, thus inhibiting scale deposits directly.

Regardless of the effectiveness of the water treatment on scale formation, it is essential to follow a viable testing program daily in order to control the chemical treatment and the blowdown. A standard kit for this purpose is shown in Fig. 8-1.

The testing procedures to be carried out, depending on plant conditions, should include checking the following constituents of the boiler water:

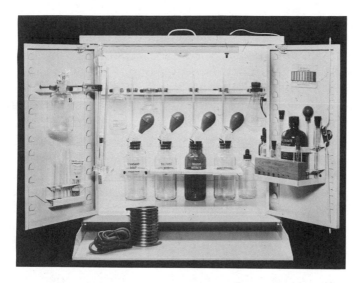

Figure 8-1. Standard procedures test kit. *(Courtesy of Carbon Solvents Laboratories, Warren, N.J.)*

Hardness: Which is the sum of the calcium and magnesium salts in the water.

Alkalinity: To prevent acidic corrosion and aid in the precipitation of calcium and magnesium.

Chlorides: To calculate rate of blowdown.

The hardness test most easily performed in the boiler room determines the precipitation of the calcium and magnesium salt with a standard soap solution. Precipitation of calcium and magnesium soaps will occur, and no lather will appear until sufficient soap has been added to precipitate all the calcium and magnesium from solution. Standard kits are available to perform this test and the result is measured in parts per million (ppm) as $CaCO_3$. The boiler water will usually show zero hardness if water treatment is correct.

The test for alkalinity is determined by titration with a standard acid solution. In this measurement the end-point is considered the point of change in the color of organic indicators, namely phenolphthalein and methyl orange. This color change shows the acid neutralizing quality of the sample being tested. The alkalinity of a boiler water should be sufficiently high to protect boiler metal from corrosion, but be sufficiently low to prevent carryover of solids with the steam. The presence of a high methyl orange alkalinity is to be avoided, because under the influence of heat in the boiler, sodium bicarbonate will break down and liberate carbon dioxide (CO_2) with the steam. Carbon dioxide is usually responsible for corrosion of steam and condensate lines. Depending upon the plant, a total alkalinity of 300 parts per million should be sufficient to precipitate the scale-forming salts and at the same time reduce the possibility of CO_2 generation. As in the case of hardness determination, standard kits are available to determine alkalinity, which is expressed in ppm as $CaCO_3$.

The chlorides of calcium, magnesium, sodium, and other elements found in water are extremely soluble and do not normally precipitate out of solution. Since no precipitation of chlorides takes place, the chlorides present in the boiler water are proportional to the total dissolved solids in the water. In boiler room practice, the rate of blowdown can be governed by controlling the chloride content of the boiler water. Determination of the chloride concentration of the feedwater and that of the boiler water provides an accurate basis for determining the percentage rate of blowdown, especially when a continuous blowdown is used. The following formula applies:

$$\frac{\text{Cl in feed water}}{\text{Cl in boiler water}} \times 100 = \% \text{ Blowdown}$$

Chloride concentration can also be used for estimating the percentage of makeup or additional water fed into the boiler. To find this percentage, determine the amount of chloride in the feedwater, the boiler water, and the returning condensate. If we assume a chloride-free condensate and there is 5 ppm in the feedwater and 10 ppm in the makeup or raw feed, then the chloride content of the feedwater is 50% of the chloride content of the makeup water so that the feedwater entering the boiler consists of 50% makeup and 50% condensate return water. The chloride test is used extensively at sea to determine condenser leakage into the steam space. For a rough check to determine the presence or absence of chlorides in the effluent from a sodium zeolite softener, a few drops of silver nitrate in a beaker of the effluent will appear as a milky cloud as the silver is precipitated as silver chloride.

In the test for chloride, the water containing the chloride is titrated with silver nitrate in the presence of potassium chromate. The chloride is precipitated as silver chloride and the end-point is reached when one additional drop of silver nitrate produces a red color. Excess sulfite in boiler water may interfere with this test, so the sulfite must be oxydized to a sulfate with the addition of hydrogen peroxide. Standard kits are available for this test.

Having a boiler that is scale-free with controlled blowdown will help in the efficient operation of the boiler room, but another source of trouble may arise from oxygen in the boiler water. Even with a well-maintained deaereator, oxygen will still enter the boiler and cause what is known as oxygen pitting. Dissolved oxygen may enter a boiler not only through raw water but in air leaks into the condensate return system, especially in heating systems that have a vacuum-assisted condensate return system. Oxygen dissolved in boiler water is released by an increase in temperature and attacks the boiler metal, causing pitting. The location of the pitting can vary with the boiler design, but is usually concentrated at the liquid level in the steam drum and usually heavily concentrated where the feedwater enters the boiler.

The most important step in eliminating oxygen from the feedwater is deaeration of the water. It is necessary to follow the mechanical deaeration with an oxygen-scavenging chemical injected into the makeup water line or into the boiler directly. Such a chemical is sodium sulfite and the oxygen-scavenging properties are illustrated by the formula:

$$2\,Na_2SO_3 + O_2 \longrightarrow 2\,Na_2SO_4$$

Sodium Sulfite plus Oxygen yields Sodium Sulfate

It takes about 10 lb of sodium sulfite to absorb one lb of oxygen, but regardless of theoretical quantities it is important to maintain residual sulfite concentrations in the boiler water, to be sure that all oxygen is scavenged. Catalyzed sodium sulfite was developed to remove oxygen from cold water, but the typical grade of commercial sodium sulfite is effective for most boilers. However, where rapid removal of oxygen from the boiler feed is necessary, the catalyzed product is recommended.

The test for the presence of sulfite in boiler water is based on titration of a sample with potassium iodide-iodate, which produces a blue color in the presence of a starch solution. Kits are available for this test.

In addition to oxygen pitting, boiler metal can be subjected to attack from an acid condition in the boiler water. If the correct alkalinity is maintained in the boiler to precipitate the calcium and magnesium salts in the form of sludge, the pH value is not necessary as a control. However, as a check on the alkalinity test it is usual when testing boiler water to report pH. For the average boiler in industrial use a pH of 10.5 is usually maintained.

The pH test was first utilized in the brewing of beer in Denmark as a control during the brewing process and was given the designation pH to stand for "power of Hydrogen." More exactly defined, it is the logarithm of the reciprocal of the hydrogen ion concentration. To simplify this, pH ranges between 0 and 14, denoting various degrees of acidity or alkalinity. On the pH scale, 7 is exactly neutral. At 0 the solution is highly acidic and as the pH number increases beyond 7 the solution becomes more alkaline.

A pH value can be determined in various ways. One is with specially designed pH paper, which is used by dipping into the solution. Depending upon color change and by comparison with set standards, the paper will indicate the approximate pH reading.

Also available are a colorimetric method that utilizes color comparison with specific standards and a glass electrode electrometric apparatus that is commercially available and utilizes a potentiometer to determine the voltage developed between two electrodes in contact with solution. Whichever method is used, determination of pH is important in boiler water control.

Another determination required to insure complete boiler water control is the amount of Total Dissolved Solids (T.D.S.) in the water. The amount of T.D.S. is used to determine blowdown rate and, depending on the temperature and pressures involved in a boiler, in conjunction with polymer chemistry, large energy savings are possible by allowing the T.D.S. to rise to previously unacceptable levels. The determination of T.D.S. can be done only in a well-equipped laboratory and involves the evaporation of the moisture from a measured sample and weighing of the remaining residue. There are, however, commercially available devices to measure T.D.S.

Having assured a clean, scale-free boiler, with no corrosion, we must now analyze other sections of the steam and condensate system that require care to prevent scale and corrosion.

Corrosion in steam and condensate lines is a problem in many plants. The costs of replacement of lines, valves and traps can be quite high and, if the problems cause interruption of production, can become excessive. Failures most frequently occur at threaded joints due to the thinning of the pipe at that location by the threads. The corrosive action results in the grooving of the pipe and pitting of the metal.

Corrosion of steam and condensate lines is caused mainly by the dissolved gases, carbon dioxide and oxygen. The source of the oxygen is the boiler feedwater or infiltration in a gravity or vacuum return system. Condensate returned under pressure is usually free of oxygen, but where the condensate is returned via a condensate pumping system, the opportunity for oxygen to be absorbed by the condensate is high. As in the elimination of oxygen from the boiler water, an efficient mechanical deaerator and the use of catalyzed sodium sulfite will eliminate most oxygen-caused corrosion.

Carbon dioxide is the usual cause of steam and return-line corrosion and is characterized by a general thinning of the pipe wall or grooving along the bottom of the pipe. The primary source of carbon dioxide is the bicarbonate and carbonate alkalinity of the make-up water. When subjected to the temperature in the boiler, the carbonates decompose and release carbon dioxide which becomes entrained with the steam.

In order to adequately control all factors in boiler water treatment, it is essential that the boiler water and condensate returns be tested a daily basis; all well-run boiler plants have a standard procedure such testing. Tests are carried out by operating personnel their regular duties.

Standard testing kits are available, as is the pr tests. The following tests are as prescribed for us cabinet (see Fig. 8-1).

Test for Al
Active P

Procedure. Pour 50 m.
Add three drops of p.
to mix contents thoroug
turn pink.

Add N/50 sulfuric acid un.
Total ml acid used × 20 .
Add three drops methyl orang

Procedure. Po

sample: again add N/50 sulfuric acid until just a drop changes the color to a reddish orange.

Total ml acid (including *P*) × 20 ppm (is the) *M* alkalinity as $CaCO_3$.

Chloride Test

Reagents Required

N/35.5 silver nitrate reagent

Potassium chromate indicator

Procedure. Use same sample that has been tested for alkalinity. Add five drops potassium chromate indicator solution. Swirl flask to mix contents thoroughly. Add silver nitrate solution until one drop changes color to red.

Test for Hardness
Soap Method

Reagents Required

Soap solution

Procedure. Pour 50 ml of the sample into shaker bottle. Run standard soap solution from burette (a few drops at a time); shake sample after each addition. When soapy lather remains on the surface for five minutes or longer, test is completed.

Total ml soap solution used minus 0.5 (lather factor) × 20 is the ppm hardness as $CaCO_3$.

Colorimetric Method (for raw water or condensate)

Reagents Required

Hardness buffer solution

(Sulfide borate)

Hardness indicator powder (AZO Dye)

Hardness titration solution (1 ml = 10 ppm as $CaCO_3$)

ur 25 ml of the sample into casserole. Add 0.5 ml hardness

buffer solution to sample and stir. Add 1 brass spoon of hardness indicator powder to sample and observe color. If color is red, hardness is present; if blue, there is no hardness present and record as zero hardness. To the red colored sample, slowly add hardness titration solution from the burette, stirring the sample until a permanent blue color remains. This is the end-point.

Total ml required × 10 is the ppm hardness as calcium carbonate.

High Phosphate Test

Reagents Required

Molybdate reagent

Dry stannous chloride

Procedure. Filter boiler water until clear, through filter paper such as Whatman #5. (It is important that no suspended matter be present in the filter sample.)

Pour filtered water into mixing tube up to 5-ml graduate mark. Add molybdate solution to 17.5 mark and mix. Add 1 dipper dry stannous chloride to tube and mix. Let stand 2 minutes. Match colors with slide comparator. The values on the slide represent ppm of phosphate as PO_4.

Sulfide Test

Reagents Required

N/40 potassium iodide-iodate

Acid starch indicator

Procedure. The sample of boiler water should preferably be obtained from a cooling coil. If this is not available, the sample should be taken in a pint bottle, filling the bottle completely and inserting the stopper or cork. The sample should then be cooled, in the bottle, to room temperature before proceeding with the test. Measure 50 ml boiler water and pour into casserole. Add 2 dippers acid starch indicator. Stir and titrate with standard potassium iodide-iodate solution from burette until a permanent blue color develops in the sample. This is to be taken as the end-point.

Calculation of Results. Using a 50-ml sample, the sulfite in parts per million as SO_3 is equal to the ml potassium iodide-iodate solution required $\times$ 20.

Test for pH of Raw or Condensate Water
(Colorimetric)

Reagent Required

Phenol red indicator
pH = 6.8–8.4

Procedure. Rinse the 5-ml test tube several times with water to be tested, then fill to the 5-ml mark. Add 0.5 ml phenol red indicator, replace the stopper and mix by inverting. Place the tube into the comparator block. Record pH number closest to matching.

EXTERNAL TREATMENT

The methods of boiler water treatment discussed so far have concentrated on the internal treatment of the water after it has been fed to the boiler. To reduce the need for internal treatment with a reduction in the use of chemicals, equipment has been developed to reduce the scale-forming properties of raw water make-up. Water contains chemicals in suspension and solution, and the particulate matter in suspension can be removed by filtration and the chemicals in solution removed by chemical and electrolytic processes.

FILTRATION

When the primary source of make-up water for a boiler is directly from a municipal system, the water is delivered as potable and should not require filtration. In many instances the source of water for make-up is a river or lake, the water being pumped directly to the boiler room; this water is contaminated with organics and inorganics in suspension as well as in solution.

Removing suspended solids by filtration is primarily a mechanical action. The solids will not pass through the filter media because of their size. Many substances are employed to filter industrial water supplies, including sand, charcoal, marble chips, diatomaceous earth, and many others.

Installations filtering cold boiler feedwater generally use sand or charcoal. Sand is the most widely used filter media for the filtration of cold water, and the right grade and size of sand and depth of bed must be carefully chosen depending on the design of the filter. The size of sand varies with the temperature of the water, the rates at which the filter is to function, the depth of the bed, and quality of the influent.

Charcoal or granulated carbon has been used for filtration of water for many years. The material possesses desirable characteristics for the filtration of water and under some conditions has the ability to entrap certain gases and chemicals. With the increasing pollution of lakes and rivers, its use is rapidly increasing.

Types of Filters

A commonly used mechanical filter is the gravity type. Constructed of concrete or steel, either rectangular or circular, the filter lends itself to the filtration of large quantities of water and is in general use by municipal water authorities and others engaged in the processing of water for general distribution. Sometimes preceded by floculation or sedimentation systems to enlarge the particle size of suspended solids, they can occupy a considerable ground area.

In the bottom of the unit is an underdrain system that collects the filtered water and also acts as the entry for backwash water. The underdrain system is buried in a bed of graded stone and gravel covered with a sand bed 24 to 36 in. deep. Figure 8-2 shows a cross-sectional view through a unit.

Water enters at the top and passes down through the sand bed into the underdrain system and then into the filtered water basin. When the unit is to be cleaned, the filtering process is stopped and filtered water is pumped into the underdrain system and up through the bed, carrying the solid material into waste water troughs and to a sewer system.

When using sand as the filtering agent, areas of the surface develop an impervious coating mixed with mud balls. This condition reduces the filters effectiveness, causing high pressure drop through the unit and channeling of the bed. When high pressure drop and channeling occur in a filter it is necessary to remove the top 2 or 3

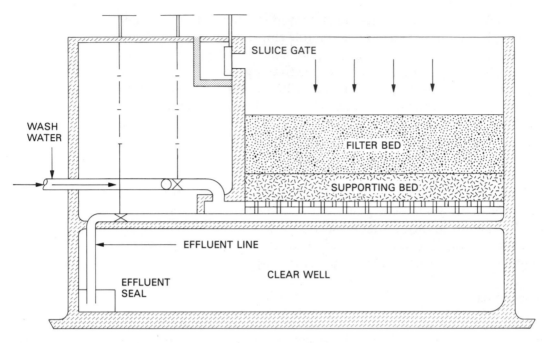

Figure 8-2. Cross section through gravity filter.

inches of the bed, taking away the accumulation of fine silt, mud, and mud balls. This condition can develop with other filter media such as charcoal or granulated carbon. Ignoring high pressure drop and channeling can result in the upsetting of the bed and loss of any effective filtration.

As with all operating equipment, performance parameters should be established when the equipment is first placed in service and any deviation from these parameters investigated.

Backwashing

The term used for cleaning the filter bed is backwashing. In most systems the backwashing is done at regular intervals of 8 to 24 hours depending on the rate of flow through the filters and the volume of suspended solids in the raw water. Backwashing is done by passing filtered water upward through the sand bed at rates from 10 to 30 gpm/ft^2 of bed area. The rate of flow should be sufficient to expand the bed from 50% to 100%.

It is important to admit the backwash water slowly, gradually increasing the flow to the desired rate. Violent admission of a reverse flow of water to the bed can result in the gravel base becoming upset. This will result in poor backwashing, poor filtration, and other operating difficulties.

Filtering operation will vary widely, depending on the time of the year and source of the raw water. Daily records of operation, covering volumes of water used and times of backwashing, will assist in the competent operation of a filter plant.

Pressure filters are essentially the same as the gravity-type filter except that the unit is an enclosed tank and operates at high pressure (Fig. 8-3). Pressure filters have found favor in industrial practice because they can be installed in line as part of a water system and do not require the large land areas of the gravity-type filter.

The pressures involved (50 to 100+ psig) require that the pressure filter be made of steel. There are two types, vertical and horizontal; the vertical is used for relatively low volume flows per unit and the horizontal when higher flows are required. The vertical units range in size from 5 to 8 ft in diameter with a 6-ft height and the horizontal units range from 8 to 10 ft in diameter and up to 25 ft

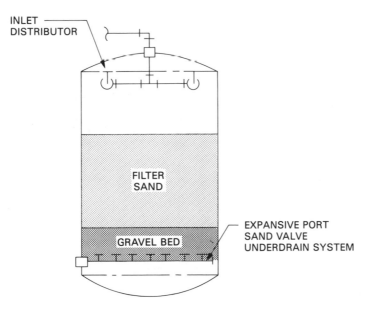

Figure 8-3. Cross-sectional view of pressure filter. (*Courtesy of Hungerford & Terry, Inc., Clayton, N.J.*)

long. Figure 8-4 illustrates the piping arrangement of a typical vertical pressure filter.

In the operation of a pressure filter the water enters at the top, passing down through the bed and out at the underdrain system, the same as in a gravity filter, but the system is under pressure at all times.

The considerations given to the selection of the filtering media are the same as those for gravity systems.

Operation

As in the operation of a boiler, the efficiency with which a filtering system is operated is directly related to the care taken by the operator in charge of the process. The filtering of water is a simple process, but the operator should have a knowledge of the principles involved.

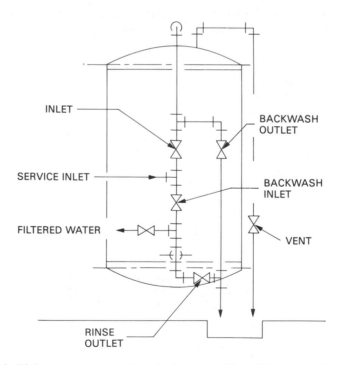

INLET

BACKWASH OUTLET

SERVICE INLET

BACKWASH INLET

FILTERED WATER

VENT

RINSE OUTLET

Figure 8-4. Piping arrangement of vertical pressure filter. (*Courtesy of Hungerford & Terry, Inc., Clayton, N.J.*)

The primary object is to remove suspended solids. This requires:

(1) Maintenance of the filter media
(2) Filters in good repair
(3) Regular and thorough cleaning of the bed
(4) Keeping a record of all operations

The filter medium is the most important part of a filter operation and regular inspection is required. With a gravity filter, a view of the top of the bed is readily available, but with a pressure filter it is necessary to stop the operation, open the unit, and inspect the exposed surface. In systems requiring continuous supplies of filtered water it is usual to install multiple units with isolation valving to allow the shutdown of individual units for maintenance service. Close attention to pressure drops through each unit will advise the operator when cleaning has become necessary.

No water purification system can be controlled without daily tests of the quality of the raw and treated water, the quantity of water treated, rates of flow, the quantity of water used for backwash, and the time between backwash. These parameters of operation, if known to the astute operator, provide the basis for the successful filtration of water.

Deaeration

The addition of sodium sulfite to the chemical feed controls oxygen content in a boiler. The use of a deaerator can reduce the amount of noncondensable gases entering the boiler. The gases of oxygen and carbon dioxide when dissolved in water accelerate its corrosive properties and high temperatures increase the rate of attack. Water will dissolve oxygen, carbon dioxide, and other gases from the atmosphere and will absorb a gas up to the point of saturation at specific temperatures and pressures.

The amount of gas dissolved by water depends on other constituents already dissolved in it. Condensate or distilled water will absorb more oxygen than will a soft water. Soft water will absorb more gases than will a hard water. By comparison, sea water will absorb a smaller amount of oxygen than fresh water, temperature and pressure being the same.

To degassify boiler feed water, an open or closed deaerator is used. In an open heater, by elevating the temperature of the feedwater to 212°F the amount of dissolved oxygen can be reduced to 0.5 ppm. These heaters are designed to heat the water by passing the water down a series of trays that provide a sheeting pattern to the water. A counterflow of steam through the water raises the temperature of the water, thus releasing the gases, which are then expelled to the atmosphere.

Deaerating heaters or deaerators provide for almost complete removal of all noncondensable gases in the returned condensate and any additional make-up water (Figure 8-5). Similar to open heaters, the units are designed with multiple-tray systems to ensure the complete removal of any dissolved gas. The counterflow system is employed with enough contact time between the water and steam flow

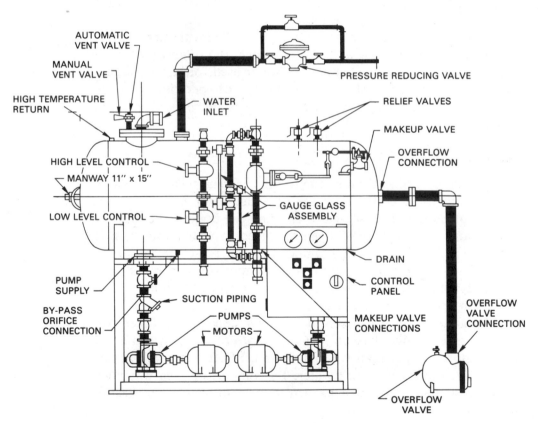

Figure 8-5. Spray-type deaerating heater. (*Courtesy of Cleaver-Brooks, Inc., Milwaukee.*)

to provide the needed efficiency. Fitted with a vent condenser in order to condense any moisture passing through the vents, these units provide high thermal efficiency. Within these units the temperature of the feedwater can be raised to 225° to 230°F, providing the heater is located in excess of 35 ft above the boiler feed pump inlet in order to prevent vaporizing of the liquid at the pump inlet and consequent vapor binding and cavitation.

SOFTENING AND DEMINERALIZATION

The removal of the calcium and magnesium from boiler make-up water can be accomplished with the use of a unit called a "brine softener" (Figure 8-6). This unit uses ion-exchange materials to exchange the calcium and magnesium ions with sodium ions. The ion-exchange material is called "zeolite" and is made of processed natural clay or synthesized material.

Sodium zeolite softening consists of passing water through a bed of material, removing the calcium and magnesium ions and replacing these ions with sodium. The exchange takes place rapidly, and calcium and magnesium will be removed from any of their salts in solution.

The typical reaction that takes place, diagrammed for the chlorides only, is:

$$Na_2Z + CaCl_2 \rightleftharpoons CaZ + 2NaCl$$

Sodium zeolite	Calcium chloride		Calcium zeolite	Sodium chloride

$$NaZ + MgZ \rightleftharpoons MgZ + 2NaCl$$

Sodium zeolite	Magnesium chloride		Magnesium zeolite	Sodium chloride

The reactions that takes place during the softening process as diagrammed for the chlorides of calcium and magnesium are the same for all forms of calcium and magnesium.

The calcium and magnesium pass through the fresh bed of sodium zeolite and are converted to the sodium form. As the flow of hard water continues, more of the zeolite becomes exchanged until the calcium and magnesium salts break through the bed, introducing

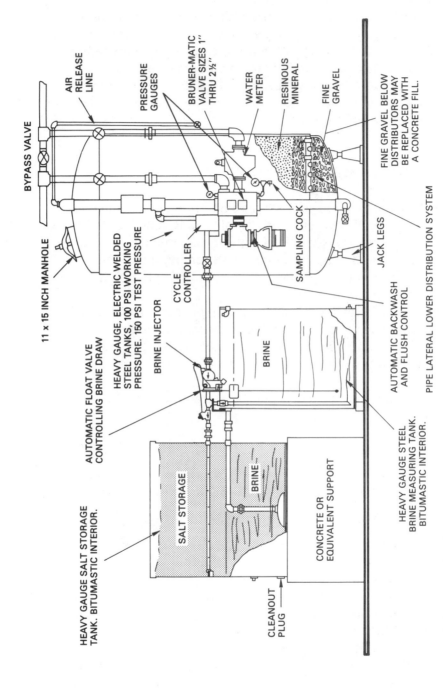

Figure 8-6. Pressure zeolite softener showing brine system, mineral bed, and underdrains. (*Courtesy of CSL Water Treatment, Inc., Warren, N.J.*)

The following labels appear on the figure:

BYPASS VALVE

AIR RELEASE LINE

PRESSURE GAUGES

BRUNER-MATIC VALVE SIZES 1" THRU 2½"

WATER METER

RESINOUS MINERAL

FINE GRAVEL

FINE GRAVEL BELOW DISTRIBUTORS MAY BE REPLACED WITH A CONCRETE FILL.

11 x 15 INCH MANHOLE

HEAVY GAUGE SALT STORAGE TANK. BITUMASTIC INTERIOR.

AUTOMATIC FLOAT VALVE CONTROLLING BRINE DRAW

HEAVY GAUGE, ELECTRIC WELDED STEEL TANKS, 100 PSI WORKING PRESSURE. 150 PSI TEST PRESSURE

BRINE INJECTOR

CYCLE CONTROLLER

SAMPLING COCK

SALT STORAGE

BRINE

BRINE

CLEANOUT PLUG

CONCRETE OR EQUIVALENT SUPPORT

HEAVY GAUGE STEEL BRINE MEASURING TANK. BITUMASTIC INTERIOR.

AUTOMATIC BACKWASH AND FLUSH CONTROL

JACK LEGS

PIPE LATERAL LOWER DISTRIBUTION SYSTEM

hardness into the effluent. The run of the softener is ended and the unit must be regenerated.

Regeneration of the zeolite is done by treating the bed with a brine solution. The reverse chemical reaction takes place, and the calcium and magnesium ions in the grains of the resin bed are released and washed to waste. The sodium is taken up by the zeolite resin and is once again capable of softening water.

DEMINERALIZATION

Softening water by ion exchange in the sodium cycle does not reduce the total dissolved solids; there is in fact a slight increase. The anions, sulfates, chlorides, and nitrates pass through the unit combined with sodium.

The development of ion-exchange materials capable of operating in the hydrogen cycle has made the removal of the sodium possible. To do this, the material is regenerated by sulfuric acid or hydrochloric acid. The chemical reactions that take place in the hydrogen exchanger for the chlorides of calcium and magnesium are as follows:

Removal

$$CaCl_2 \;+\; H_2Z \longrightarrow 2HCl \;+\; CaZ$$

$$MgCl_2 \;+\; H_2Z \longrightarrow 2HCl \;+\; MgZ$$

Regeneration

$$CaZ \;+\; 2HCl \longrightarrow 2H_2Z \;+\; CaCl_2$$

$$MgZ \;+\; 2HCl \longrightarrow H_2Z \;+\; MgCl_2$$

The calcium and magnesium chlorides in the untreated water combine with hydrogen ions; the calcium and magnesium remain in the bed combining with the ion-exchange material. The treated water contains the free mineral acid HCl, which must be removed.

This is done by either adding caustic soda to neutralize the acid or by mixing the acidic effluent with effluent from a sodium unit that contains sodium bicarbonate. By blending, the desired alkalinity can be reached. Table 8-1 shows the results of blending water from

TABLE 8-1
Results of a Water Blend from Sodium and Hydrogen Units

	PPM	PPM	PPM	Blend 25% Na, 75% H$_2$
Cations (as CaCO$_3$)				
Total hardness	157	2	2	2
Calcium	99	1	1	1
Magnesium	58	1	1	1
Sodium	16	171	2	44
Hydrogen	0	0	32	0
Anions (as CaCO$_3$)				
Bicarbonate	137	137	0	10
Chloride	16	16	16	16
Sulfate	20	20	20	20
Other				
Carbon dioxide	2	2	123	5
Silica	3	3	3	3
Total solids	176	176	39	49

sodium and hydrogen units, and Figure 8-7 is a schematic view of the blending operation.

MIXED-BED DEMINERALIZATION

To obtain the same results in a smaller space and with less capital outlay, the use of mixed-bed demineralizers is now a most economic solution to providing pretreatment for boiler feed water. Figure 8-8 shows a cross-sectional view through a mixed-bed deionizer. The resins are shown separated prior to regeneration. This separation is achieved by backwashing the unit. A difference in the specific gravity of the anion and cation resins allows the two resins to separate.

After separation, a dilute sodium hydroxide regenerant is injected into the caustic inlet valve (Figure 8-9). The solution passes downward through the anion bed, removing contaminants from the resin and going to waste at the interface distributor outlet.

After rinsing the anion resin, a hydrochloric acid regenerant is injected at the acid inlet, passing up through the cation bed and to

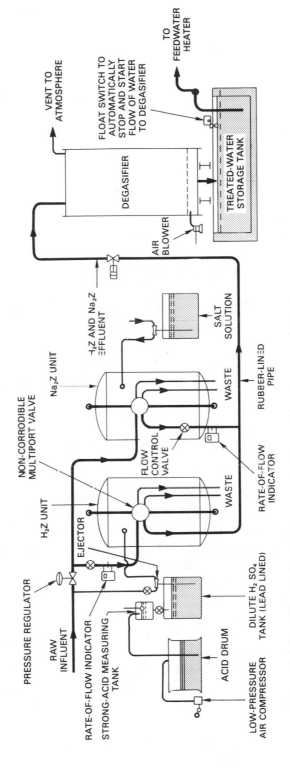

Figure 8-7. Blending effluents from hydrogen and sodium units. (*Courtesy of CSL Water Treatment, Inc., Warren, N.J.*)

Labels within figure:

PRESSURE REGULATOR

RAW INFLUENT

RATE-OF-FLOW INDICATOR

STRONG-ACID MEASURING TANK

H₂Z UNIT

EJECTOR

NON-CORRODIBLE MULTIPORT VALVE

Na₂Z UNIT

H₂Z AND Na₂Z EFFLUENT

VENT TO ATMOSPHERE

FLOAT SWITCH TO AUTOMATICALLY STOP AND START FLOW OF WATER TO DEGASIFIER

TO FEEDWATER HEATER

DEGASIFIER

TREATED-WATER STORAGE TANK

AIR BLOWER

SALT SOLUTION

WASTE

RUBBER-LINED PIPE

FLOW CONTROL VALVE

RATE-OF-FLOW INDICATOR

WASTE

DILUTE H₂SO₄ TANK (LEAD LINED)

ACID DRUM

LOW-PRESSURE AIR COMPRESSOR

213

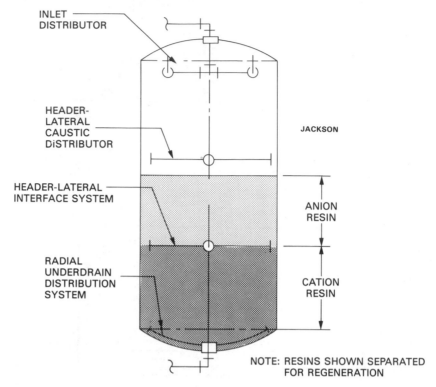

INLET
DISTRIBUTOR

HEADER-
LATERAL
CAUSTIC
DiSTRIBUTOR

JACKSON

HEADER-LATERAL
INTERFACE SYSTEM

ANION
RESIN

RADIAL
UNDERDRAIN
DISTRIBUTION
SYSTEM

CATION
RESIN

NOTE: RESINS SHOWN SEPARATED
FOR REGENERATION

Figure 8-8. Cross-sectional view of mixed-bed demineralizer. (*Courtesy of Hungerford & Terry, Inc., Clayton, N.J.*)

waste at the interface outlet. The cation resin is then rinsed. After regeneration the entire unit is air mixed and a final rinse removes any regenerant remaining in the bed. The location of the valves allows the operator to inject both acid and caustic regenerant at the same time. This action results in a neutralizing of the water to waste and a reduction in the time required for regeneration.

The quality of water produced by mixed-bed demineralization can be the equivalent of distilled water without the large amount of energy expenditure needed by such units. The quality of demineralized water is measured in specific resistance or specific conductance. Table 8-2 illustrates water quality in microsiemens per centimeter and ohms per centimeter contrasted with the dissolved solids in parts per million as sodium chloride.

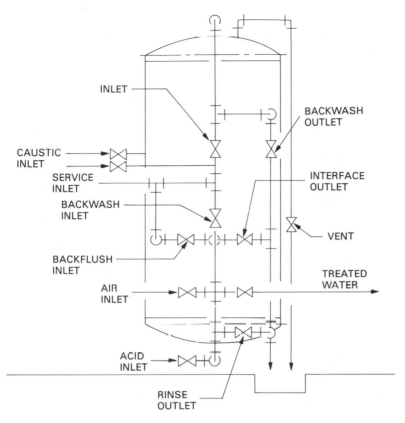

Figure 8-9. Valve arrangement on a mixed-bed demineralizer. (*Courtesy of Hungerford & Terry, Inc., Clayton, N.J.*)

TABLE 8-2
Conversion Chart

Specific Conductance μs/cm	Specific Conductance Ω/cm	NaCl (ppm)
0.038	26,000,000	None
0.056	18,000,000	0.03
0.1	10,000,000	0.04
0.2	5,000,000	0.13
1	1,000,000	0.50
2	500,000	1.0

TABLE 8-2 (cont.)

Specific Conductance $\mu s/cm$	Specific Conductance Ω/cm	NaCl (ppm)
4	250,000	2.0
6	166,666	3.0
8	125,000	4.0
10	100,000	5.0
20	50,000	10
30	33,333	15
50	20,000	25
100	10,000	50
200	5,000	100
500	2,000	250
1,000	1,000	500
2,000	500	1,000
5,000	200	2,500

EXERCISES

1. How do impurities and chemical compounds get into boiler feed water?

2. What chemicals are the source of scale in boilers?

3. Why does boiler scale reduce the efficiency of a boiler?

4. What pH would you carry in boiler water and why?

5. If salts are to be precipitated without raising the alkalinity, what chemical compound is used?

6. What causes "hardness" in boiler water?

7. What is the best method to calculate the blowdown rate?

8. What causes pitting of the boiler shell and what methods are used to prevent pitting?

9. What causes corrosion in condensate lines?

10. What is methyl orange? What is phenolphthalein?

11. When is it desirable to use pressure filters instead of gravity filters?

12. How is the oxygen-carrying capacity of water affected by its chemical constituents?

13. In your own words, describe the process of demineralization.

14. What effect do mud balls and silt have on the operation of a sand filter?

MULTIPLE CHOICE QUESTIONS

1. Water containing a high proportion of chemicals is called:
   a. Hard
   b. Soft
   c. Treated

2. CO^2 is the chemical symbol for:
   a. Carbon monoxide
   b. Carbon dioxide
   c. Carbonic gas

3. Scale on a boiler surface will result in:
   a. Wasted water
   b. Wasted heat
   c. Wasted steam

4. Scale is formed in a boiler by precipitated:
   a. Chemicals
   b. Carbonates
   c. Salts

5. Chlorides help in water treatment:
   a. By preventing scale
   b. By calculating rate of blowdown
   c. Neither a nor b

6. To remove suspended solids from water you need:
   a. Filtration
   b. Precipitation
   c. Atomization

7. The soap solution test is recorded in:
   a. ccm
   b. ppm
   c. mg/l

8. Calcium and magnesium can be removed with:
   a. Sodium carbonate
   b. Precipitation
   c. Sodium zeolite

9. A high pH shows the water to be:
   a. Neutral
   b. Alkaline
   c. Acidic

10. Soft water will absorb more gas than:
    a. Hard water
    b. Distilled water
    c. Both a and b

Resistance Pyrometer

The resistance of an electrical conductor changes when the temperature of that conductor changes, and this knowledge is used in the design of the resistance pyrometer. This instrument is suitable for measuring temperatures from subzero to 2000°F with accuracy within 0.5%. A resistor wire of platinum, a nickel alloy, or copper is embedded in an insulator. The electrical resistance of the wire varies precisely with the temperature, and this change in resistance can be accurately measured. A heavy-wall stainless-steel sheath encloses the sensing portion of the element to provide protection from a hostile environment (Fig. 9-3).

The recording and measurement of temperatures is an essential part of boiler operation and all the sensing devices discussed so far are utilized in a boiler room. The ability to transmit the temperature by

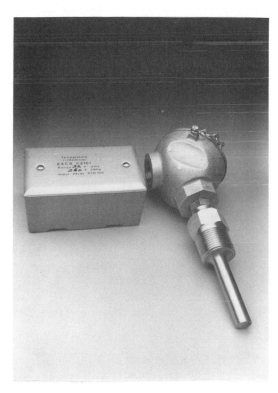

Figure 9-3. Resistance pyrometer. (*Courtesy Leslie Co., Parsippany, N.J.*)

9

Instruments and Controls

In previous chapters we discussed the need to measure and control the temperatures, pressures, rate of flow, quantity, and quality of the elements and materials that combine to produce steam in a controlled environment. Boiler room instrumentation is constructed and designed to provide the measurement and control. Dictionaries list several definitions of instruments or instrumentation but the one that concerns us is "devices for the measurement and control of complex systems."

TEMPERATURE MEASUREMENT

The thermometers of Fahrenheit and Celsius provided the first measurement of temperature that was reproducible, and the same glass tube with an element of mercury is in use today to measure temperatures within the range of those instruments.

The development of devices capable of generating heat far beyond the range of the thermometer made it necessary for the scientific and engineering communities to also develop a means of recording these higher temperatures, that is, a recorder or measuring device that would operate in a hostile environment.

Bimetallic Element

Utilizing the principles of the unequal expansion and contraction of dissimilar metals, recorders and indicators have been devised using a bimetallic element as the actuating source. The dial thermometer uses this principle to rotate a needle around a concentric dial to indicate the temperature. This element is enclosed in a strongly constructed protective well.

Filled Systems

When it is necessary to read a temperature at a point remote from the source of heat, a filled system is used. A filled system thermometer consists of an indicating instrument with its pointer actuated by a Bourdon tube connected by capillary tubing to a filled bulb. The bulb can be filled with a liquid or gas that will respond to a change in temperature; the pressure in the enclosed system rises when the temperature increases and falls on a decrease in temperature. The pressure differential actuates the Bourdon tube and through linkage moves a pointer over a scale indicating the temperature (Fig. 9-1).

Thermocouple

In addition to the filled bulb and capillary tubing, the use of the thermocouple can provide temperature readings at a distance. A thermocouple is a pair of conductors of dissimilar metals joined together at two points so that an electromotive force proportional to the temperature difference is generated if the two points are at different temperatures (Fig. 9-2). The scale at the voltmeter can be graduated to read the temperature in degrees.

The metals used for a thermocouple can be divided into two groups, base metal and noble metal. The base metals are iron, nichrome, or constantan. Constantan is an alloy of copper and nickel. Base-metal thermocouples are used for the lower range of temperatures up to 1600°F. The noble metal used is platinum, one wire being 100% platinum and the other 90% platinum and 10% rhodium. The combination of metals chosen depends on the type of service required and the temperatures involved.

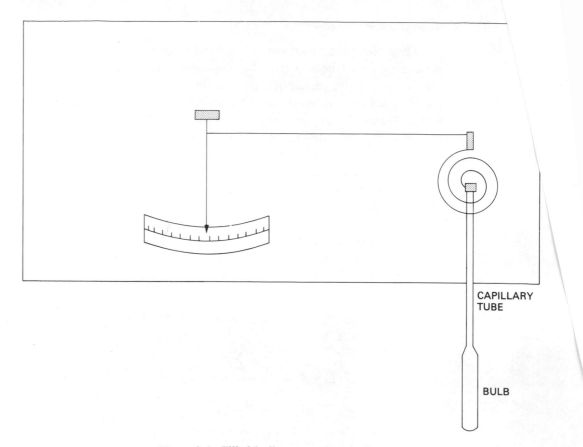

Figure 9-1. Filled bulb temperature indicator.

CAPILLARY TUBE

BULB

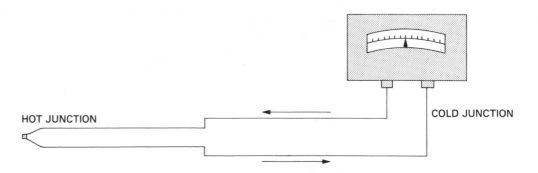

HOT JUNCTION

COLD JUNCTION

Figure 9-2. Thermocouple circuit.

capillary tube or electrical wire also gives us the capability to control operations within temperature limitations, remote from the temperature source. This capability is used for the control of hot-water boilers, fuel oil temperatures, stack temperatures and feedwater temperatures, allowing the operator to preset conditions for maximum operating safety and efficiency.

PRESSURE MEASUREMENT

The operation of a boiler plant requires the measurement and recording of the numerous pressure points throughout the system. The pressures can be defined in three categories: positive pressure, vacuum, and absolute pressure.

The measurement of pressure is as a differential in reference to barometric pressure. Vacuum is measured below atmospheric pressure, and absolute pressure takes into consideration the variations in atmospheric pressure as measured by a barometer.

The operation of a pressure gauge is based on one of the following: the Bourdon tube, the manometer, or the diaphragm. The manometer is the simplest of pressure gauges and is particularly adapted to the measurement of very low pressures and vacuum. Typical boiler room uses for the manometer are the measurement of draft through a furnace or air and gas pressures measured in ounces per square inch or inches of water. Mercury-filled and diaphragm-operated units are available for installation in boiler room control panels. However, all units work on the same principle; the pressure is determined by the difference in the levels of a liquid in two legs of a U-tube (Figure 9-4).

The Bourdon tube pressure gauge depends for its action on the change in curvature in a curved metallic tube when the pressures are varied. This type of gauge is used to measure pressures below atmospheric and up to 100,000 lb. Graduated on a concentric scale in pounds per square inch for those pressures above atmospheric and in inches of mercury for pressures below atmospheric, the Bourdon tube gauge meets all standards of accuracy and repeatability required of such an instrument. For high pressures the tube is constructed of steel or stainless steel sufficiently rugged to withstand the service of fluctuating pressure. For lower pressures the tube can be constructed of phosphor bronze.

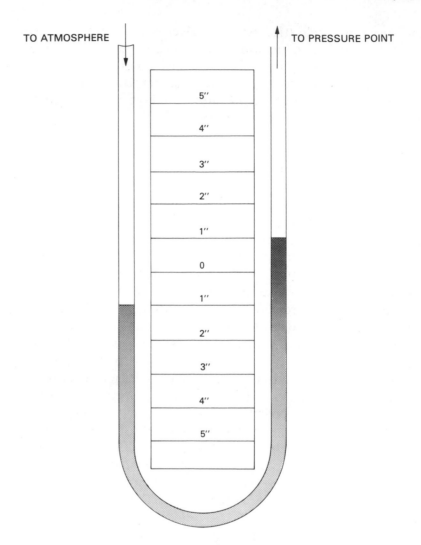

TO ATMOSPHERE   TO PRESSURE POINT

5″

4″

3″

2″

1″

0

1″

2″

3″

4″

5″

Figure 9-4. Manometer.

When pressure is applied to the tube it tends to straighten out, and a link and pinion attached to a quadrant gear assembly actuates a pointer. The pointer moves through an arc of about 270° from pin to pin (Figure 9-5). When a gauge is constructed to read both pressure and vacuum, it is called a "compound pressure gauge."

The diaphragm-type pressure gauge is actuated by the pressure causing the diaphragm to bulge toward the area of least pressure. This movement is conveyed to the indicating pointer via a connecting

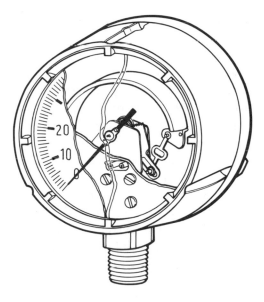

Figure 9-5. Pressure gauge. (*Courtesy of WIKA Instrument Corporation, Hauppauge, N.Y.*)

strut and a gear and pinion. The use of the diaphragm isolates the material or element causing the pressure from the operating mechanism. If a suitable material for the diaphragm is used, this type of gauge can be used on a chemical feed line to a boiler, since the highly caustic chemicals will be isolated from the mechanism.

FLOW MEASUREMENT

In a boiler room, water, oils, gases, and steam are continuously flowing throughout a complex system of piping and valves. For the control and measurement of these flows, various measuring and control devices are used. The volume of any flow depends on the size of the pipe and the pressure inside the pipe. Another factor would be the differential between the initial pressure and the terminal pressure.

Steam Flow

A majority of steam-generating systems use a differential pressure device to measure steam flow. This consists of an orifice plate between two flanges. The orifice creates a difference in pressure between the upstream and downstream sides of the orifice plate. This pressure difference is measured by a differential pressure meter and is converted to an electronic signal. The electronic signal is conveyed to the recording device.

As energy management has received greater attention in recent years, combinations of flow and Btu measuring devices have been devised to give the operating engineer or energy manager a more accurate picture of energy use. Vortex shedding meters, direct reading meters, ultrasonic, and turbine meters and their combinations are utilized to record energy flow.

Pitot tubes have been used in flow measurement for many years. The system in Figure 9-6 utilizes an averaging pitot tube to provide accurate flow-rate sensing. The system can measure steam flow by sampling pressure and differential pressure and then calculating steam flow. A system of this type provides the energy manager with a reliable means of measuring steam flow in a complex system of piping.

Btu Meters

Utilizing temperature and flow Btu meters measure the temperatures on the supply side and the return side. A turbine measures the flow and an electronic calculating assembly multiplies the tempera-

SATURATED STEAM APPLICATION

CALCULATES INSTANTANEOUS AND TOTALIZED MASS FLOW RATE
(LB_m/HR)

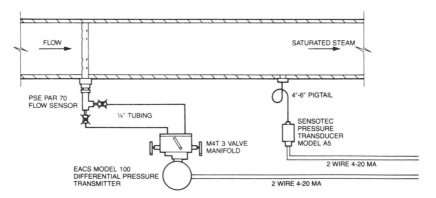

Figure 9-6. The Commander 5000 measures steam flow. (*Courtesy of Leslie Co., Parsippany, N.J.*)

ture change by the flow. The Btu's used are shown on a counter (Figure 9-7).

Liquid Flow

Liquid flow in a boiler room is either water or oil. The measurement points for water are usually the make-up line to the feed heater or the feedwater line into the boiler. The measurement of oil flow to the burner is necessary in order to calculate the efficiency of the boiler and account for oil usage in the plant. A true accounting of all oil on site assures the operator that there are no unseen losses from the leakage of tanks or pipe lines. Most modern meters for measuring liquid flow are of the rotary volumetric type. A rotating chamber transmits the flow to a gear train that turns a numbered counter, which displays the total flow in gallons.

In measuring the flow volume, either for oil, water, or gas, for

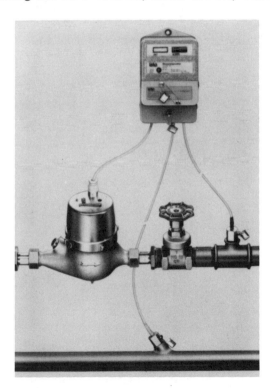

Figure 9-7. Btu meter with separate flow meter. (*Courtesy of ISTA Energy Systems Corp., Roselle, N.J.*)

complete accuracy, temperature compensation must be made. Gas meters are sometimes temperature compensated. Oil flow can be adjusted by using Table 3-3 to reduce actual oil volume measurement to the standard 60°F.

The reading of the counter at regular intervals and recording the reading will provide a continuous source of data for measuring the efficiency of the steam system (Figs. 9-8 and 9-9).

BOILER WATER LEVEL CONTROL

The continuous production of steam requires the continuous addition of water to the boiler and the maintenance of a constant fixed level of water in the boiler to cover those parts of the boiler exposed to the heat of the fire and provide sufficient space for the accumulation of steam. The elements that act to change the boiler water level are the steam load, water flow rate, steam pressure, and rate of firing. Each of these elements can be incorporated in a control system.

Figure 9-8. Oil flow meter. (*Courtesy of ISTA Energy Systems Corp., Roselle, N.J.*)

Figure 9-9. Water flow meter. (*Courtesy of ISTA Energy Systems Corp., Roselle, N.J.*)

One-element Control

In a one-element control the element used is the water level. A float-operated control responds to the water level in the boiler and sends a signal to the feedwater control valve to open or close. Built into this system is a low-water cutoff switch that cuts off the fire when water level falls to an unacceptable level.

Two-element Control

A two-element control utilizes steam flow plus water level for control elements. As the steam load increases, the water level will tend to rise because of the pressure drop in the drum. The increase in steam flow is fed to the liquid-level controller to counteract the apparent rise in level. The valve position is a sum of the steam flow element and the water level.

Three-element Control

Systems utilizing three-element control can maintain constant water level regardless of demand or load fluctuation. Using the water level, steam demand, and one other element (either water flow or firing rate), valve positioning is regulated by proportional plus integrated action.

COMBUSTION CONTROL

To maintain a constant steam pressure and flow, the rate at which fuel is admitted to the furnace must be matched by the flow of steam from the boiler or, in the case of a hot-water boiler, the preset temperature of the circulating water. The steam pressure is measured and compared to a set point on a master controller that determines the settings for the fuel and the air.

Fuel flow for oil firing is controlled by an oil metering valve controlled by an oil metering cam receiving a signal from the master controller via a jack shaft (Fig. 9-10). When the fuel is gas, the controlling valve is a butterfly valve controlled by the signal from a master controller. For coal firing, the speed of a traveling grate or rotary screw feeder is varied, with various other arrangements used when burning powdered coal.

The controls common to all boilers when firing oil are:

(1) *Forced draft fan:* Provides all air for combustion of pilot fuel and main fuel and for purging the boiler fireside prior to light off.

(2) *Ignition transformer:* Provides high-voltage spark for ignition of gas pilot.

(3) *Modulating motor:* Operates rotary air dampers and fuel valves.

(4) *Programming and flame safeguard control:* Automatically programs the starting operation and safety shutdown of the boiler in conjunction with relays and limit switches.

(5) *Operating pressure control:* Stops burner operation on a rise in boiler pressure above preset limits.

(6) *Modulating pressure control:* Senses changes in the boiler pressure and transmits to a modulating motor to reduce or increase fuel flow.

(7) *Low-water cutoff:* This control responds to water level in the boiler and shuts the fire off in the event of low water.

(8) *High- and low-pressure limit switches:* These switches prevent startup of the boiler if the oil, air, or gas pressure is either too high or too low compared to a preset range.

When gas is used as a fuel, a series of controls for the safe burning of the gas precede the burner. The typical gas train (Fig. 9-11) incorporates the following:

(1) *Gas pilot valve:* A solenoid valve that opens during the ignition period to admit gas to the pilot. It closes when the main flame is established.

(2) *Gas pilot shutoff cock:* For manually opening and closing the gas supply to the pilot valve.

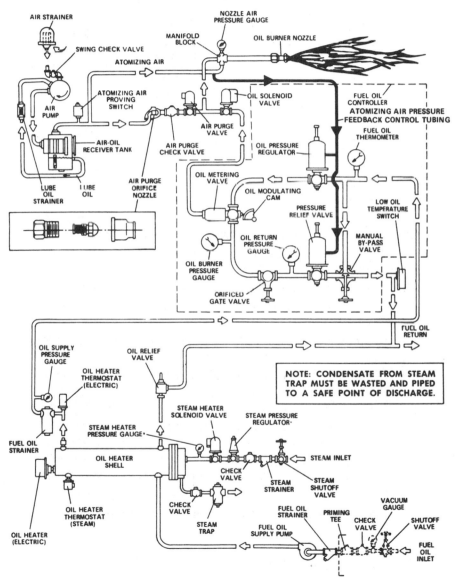

Figure 9-10. Schematic diagram of heavy oil firing. (*Courtesy of Cleaver-Brooks, Inc., Milwaukee.*)

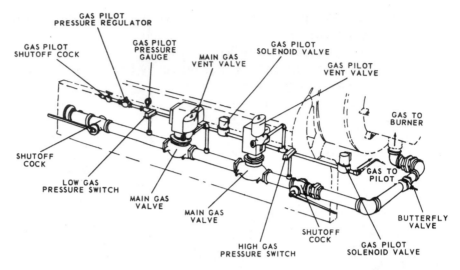

Figure 9-11. Typical gas train. (*Courtesy of Cleaver-Brooks, Inc., Milwaukee.*)

(3) *Gas pilot adjusting valve:* Regulates the size of the flame.

(4) *Gas pressure gauge:* Indicates gas pressure to the pilot.

(5) *Gas pressure-regulating valve:* Reduces gas pressure to suit the pilot's requirements.

(6) *Butterfly gas valve:* This valve is actuated by connecting linkage from the modulating cam to regulate flow to the burner.

(7) *Gas-modulating cam:* An assembly provided for adjustment of gas input to the burner.

(8) *Main gas valves:* Electronically actuated shutoff valves that open to admit gas to the burner.

(9) *Main gas vent valves:* A normally open solenoid valve installed between the two main gas valves to vent gas to the atmosphere should any be present when main gas valves close. This vent valve closes when the main valves open.

(10) *High and low gas pressure switches:* A pressure switch that is closed when the gas pressure is within preset limits. These switches usually have to be manually reset when activated.

STACK GAS MEASUREMENT

From the days of measuring efficient operation by the ''economy haze'' at the stack, through Orsat analysis to oxygen trim control,

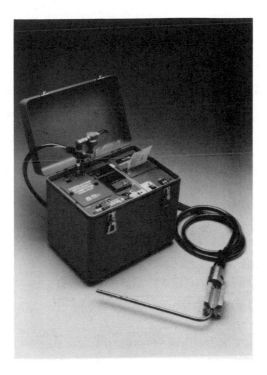

Figure 9-12. Enerac Model 942-XP. (*Courtesy of Energy Efficiency Systems, Inc., Westbury, N.Y.*)

boiler operators have always been aware that emissions from the stack represent the largest losses in boiler operation. As previously discussed, low excess oxygen in the emission is a means of burner control, but excess carbon monoxide production can be the result of too low an oxygen content in the stack gases.

To prevent excessive carbon monoxide production, the measurement of CO is now a means of control for boiler efficiency. The Enerac model 942-XP is the latest device for such measurement (Figure 9-12). This combustion efficiency computer records on a printed tape the temperature, combustibles, carbon monoxide, oxygen, carbon dioxide, and overall boiler efficiency. A systematic program of recording the temperature and the contents of the stack gases provides a positive guide for assuring boiler efficiency.

EXERCISES

1.   What is a thermometer?

2.   What principle is used in a bimetallic element?

3. If a manometer reads a negative pressure, at which point on a boiler could it be connected?

4. What is the purpose of oxygen analysis?

5. What makes automatic control an economic necessity?

6. What danger can arise from the generation of CO?

7. In what area of the boiler room is a pyrometer used?

MULTIPLE CHOICE QUESTIONS

1. Instruments in a boiler room are designed to provide:
   a. Measurement   c. Results
   b. Control   d. Both a and b

2. A factor in the design of boiler room controls is:
   a. Heat and Pressure   c. Cost of instruments
   b. Operate in a hostile environ-   d. Efficiency
      ment

3. A filled system is used to measure:
   a. Height   c. Temperature
   b. Pressure   d. Both a and b

4. In a filled system, pressure rises when:
   a. Pressure falls   c. Temperature falls
   b. Temperature rises   d. None of the above

5. A thermocouple is another name for:
   a. Two conductors   c. Two different metals
   b. Two thermometers   d. Two temperatures

6. The noble metal used in a temperature-measurement instrument is also called:
   a. Copper   c. Platinum
   b. Constantan d. Unknown

7. A resistance pyrometer is used to measure:
   a. Pressure   c. Resistance
   b. Temperature   d. Resistivity

8. A manometer is used to measure:
   a. Flows
   b. Pressure
   c. Light
   d. Heat

9. The arc through which a pressure gauge pointer moves is:
   a. 90°
   b. 360°
   c. 180°
   d. 270°

10. A compound pressure gauge will read:
    a. Two pressures
    b. High and low pressure
    c. Pressure and vacuum
    d. Both a and b

11. Measuring Btu consumption is used to record:
    a. Energy use
    b. Efficiency
    c. Escapement
    d. Vaporization factor

12. One-element feedwater control uses:
    a. Water level
    b. Steam demand
    c. Firing rate
    d. Feedwater flow

10

Managing the Power Plant Operation

The role of the power plant operator in the operation and maintenance of the boilers is primarily to ensure safety and efficiency. The operator must also manage the shift or the entire operation in the most progressive manner.

The use of steam generators reflects the need for steam and the resultant energy that steam produces.

The equipment involves a capital investment from which is expected a return, consistent with the efficient management of capital resources. In addition to the capital investment, expenses include labor and raw materials, such as fuel and water. Any losses from mismanagement of purchased materials will increase the cost of investment.

To enable the operator to run his plant efficiently, it is important to establish operating parameters at the time the system is installed. When equipment is first installed, there are specified standards of performance that the equipment must meet. A boiler will generate a certain number of pounds of steam for each pound of fuel burned, a pump will pump a specific amount of water per minute, a compressor will produce a certain quantity of air at a predetermined pressure, etc. All these capabilities should be tabulated on an equipment card with all other parameters of performance, so that these standards can be maintained throughout the life of the equipment and deviations corrected by proper maintenance.

To evaluate equipment performance, the most important record is the daily log sheet. All boiler operations should be reflected in hourly readings taken by the operator on duty. A typical log sheet, as illustrated in Fig. 10-1, records the pressures, temperatures, and flows of all the vital components of a boiler operation. Any change from normal performance is immediately apparent.

At the end of each watch, the reading should be recorded on a chart, providing a specific record of any changes. The fuel oil and water consumed should also be recorded on a monthly chart as a further check on any change requiring correction. An additional check

XYZ Co.   Date _____

Boiler # _____

Time								Remarks
Fuel Temp °F								
Fuel Press lb								
F.W. Press lb								
F.W. Temp °F								
Stack Temp °F								
Boiler Control Condition Hand/Auto								
Draft (Furnace)								
Draft (Stack)								
Fuel Oil Suction Hi/Lo								
F.O. Tk #								

Meter Readings

Fuel Oil   Water

Finish of Watch _____   _____

Start of Watch _____   _____

Total Consumed _____   _____

Boiler Blowdown at   No. Seconds each Blow

Tubes Blown at   No. Revolutions

Remarks _____

_____

_____

_____

_____

Operator _____

Figure 10-1. Boiler room log.

on performance is the cost of each thousand pounds of steam generated. This figure is established by tabulating the total cost of the monthly operation and dividing that number by the amount of steam in pounds produced.

For example:

$$\text{Cost of operation/month} \quad = \quad \$154,000$$

$$\text{Pounds of steam produced/month} = 40,000,000$$

Therefore, the cost per 1000 lb of steam is \$3.85.

A monthly calculation of this performance will serve as a guide to the efficiency of operation. Management will, of course, seek to minimize the cost per thousand pounds.

The graph of daily steam production vs. fuel and water consumption illustrated in Fig. 10-2 is a helpful chart for continuous comparison of efficiency standards. The typical chart as recorded shows the steam and fuel lines running almost parallel, any deviation could be explained by load changes and the failure of the fuel/air ratio to remain constant at all loads. The water consumption is the result of leakage in the steam system, the failure to return condensate, and blowing down of the boiler. Clearly, if all steam produced were returned to the boiler as feed, the water consumption would be a straight line, accounted for by just the blowing down of the boiler and the scavenging steam used in a deaerating feed heater. The important point to remember is that all materials consumed in the production of steam should be monitored in a logical manner.

BOILER INSPECTION

The commercial operation of any boiler is subject to regulation by at least one government agency and sometimes by several local and state authorities. The purpose of such regulation is primarily to ensure safety. A boiler under pressure and not properly maintained is an extremely dangerous device. To guard against any financial loss due to an explosion, fire or other mishap, boiler owners provide for insurance. The insurance companies require at least one annual inspection of a boiler to assure safe operation and to protect their assets.

It is the duty of operating personnel to prepare the boiler for such an inspection by doing the following:

(1) Shut off burners or clear grates of all coal.

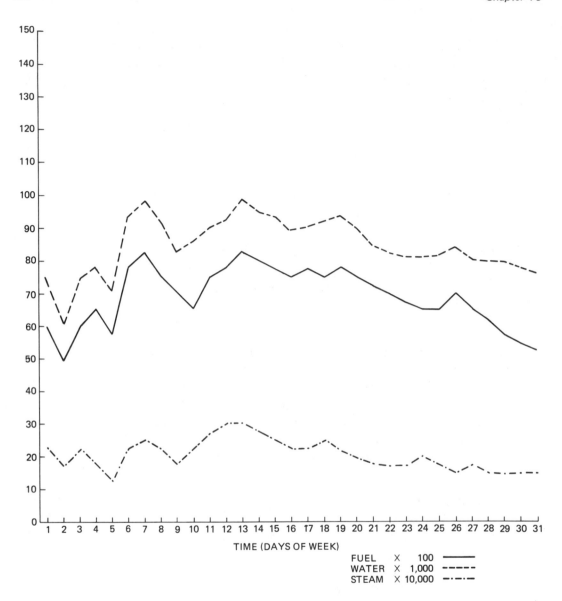

Figure 10-2. Daily steam production vs. fuel and water consumption.

(2) Allow the boiler to cool.

(3) Drain all water from the boiler and isolate the feedwater makeup valve. Open the air vent to break the vacuum in the boiler. The feed valve should be locked in the closed position and a "Danger-Do Not Open" tag placed on the

valve. If the feedwater valves have a tendency to leak, break the piping at a suitable flange to prevent any water from entering the boiler.

(4) Disconnect fuel or gas lines to the burner. Also disconnect any steam or air atomizing lines to the burner.

(5) Close the main stop valve, lock, tag it. In single-boiler operations, taking precautions against backflow into the boiler may appear overly cautious, but if you get into the habit of safe operation, you will take these necessary precautions automatically on multiboiler plants.

(6) Open up the burner front or man doors to provide access to the fireside of the boiler. Set up blowers or other means of circulation to thoroughly vent the fireside. On a horizontal return tube boiler, the entire front and rear doors can usually be opened to provide access to the tubes.

(7) Open the highest door into the waterside of the boiler. It is important to open this door first to provide a means of breaking any vacuum that has formed in the boiler during draining. Even though the air vent has been opened, if it became plugged the water would stop draining, indicating an apparently empty boiler. Opening a lower door first while water is still in the boiler risks the possibility of scalding the operator and flooding the boiler room.

(8) Open all doors to the waterside of the boiler.

(9) Clean the waterside of the boiler. Wire brush all accessible surfaces. If you are operating a water tube boiler, run an air-operated brush through the tubes, making sure they are all clear.

(10) Inspect the waterside of the boiler. Look for any sign of corrosion, pitting, or any other deficiency. Make sure the feedwater piping is clear. Check the dry-pipe for any build-up in the perforations. Make sure the chemical feed piping is clear. Examine the reinforcing ring at man holes and hand holes for any corrosion or cutting of the door surface. Check the fusible plug, if one has been installed, for tightness and ensure that the fusible filling in in good condition. If you replace the fusible plug, remember that it may be a waterside or fireside plug. Replace with the correct one.

(11) Clean the fireside of the boiler, removing all soot, ash, etc. Do not use steam or water to clean the fireside of the boiler. The moisture added will combine with sulfur in the fuel

and form sulfuric acid, a very corrosive derivative. Check all refractory and other forms of insulation. Replace or repair as necessary. Clean all fireside tubes down to the metal. Remember $\frac{1}{16}$ in. of soot or scale on the tubes will reduce the efficiency of the boiler.

CLOSING THE BOILER

After inspection, the boiler should be closed, filled with water, and put back into service. Those few words cover a broad area of operation and require elaboration.

Before closing the boiler, it is essential to examine all sections of it to ensure that all tools, brushes, rags and other cleaning materials have been removed. A rag or other foreign material that gets into a tube is a sure source of trouble.

Having inspected all areas thoroughly, start closing all waterside hand holes and man holes, replacing all gaskets with new ones and applying a mixture of steam cylinder oil and graphite to the gaskets. This will prevent the gasket from sticking to the metal when you next open up the boiler. If steam cylinder oil and graphite are not available, any of the high-temperature anti-sieze compounds available may be used. When all waterside doors are closed, begin filling the boiler. Let the boiler fill until the gauge glass is ¼ full and then make a final check of the fireside, to ensure against any leaks.

Raising Steam

Close up the fireside and start the burner, making sure that the air vent on the boiler top is open. Depending upon the size and type of the boiler, raise the temperature of the boiler very slowly, keeping the smallest fire possible until steam starts to blow from the air vent. This will ensure that all the air has been removed from the boiler water.

When steam is blowing freely from the vent, close the vent. During the period of warmup, tighten all hand and man hole doors every half hour. As the boiler heats up, the length of the bolts increases and a door that was tight when cold will loosen. The periodic tightening of the doors should be continued until a full head of steam has been raised on the boiler.

While raising steam it is also good practice to check the low-water cutout and the flame failure relays to make sure they work before you have a full head of steam.

CUTTING IN A BOILER

The practice of cutting in a boiler on a multiboiler plant is relatively simple, but some precautions are necessary.

Where several boilers feed into a main header it is usual to have a nonreturn main stop on each boiler. As you are raising steam, open the main stop full and open the header valve full. The main stop and connecting pipe to the main header are fitted with drains blowing to the atmosphere. These drains should be fully open to remove any water from the line. With both stop valves open as the pressure rises on the boiler being cut in, the nonreturn valve will automatically open when the boiler pressure slightly exceeds the main header pressure, allowing the boiler to start delivering steam to the system.

On small boiler operations the main stop valve is usually a gate valve. This valve may be opened before firing the boiler so that as steam is produced it will automatically fill the system.

If steam is raised with the main stop valve closed, a globe valve bypass should be fitted around the gate valve in order to balance the pressure on both sides of the gate to make it easier to open.

In all changes of operation of a boiler, be methodical and deliberate in all your actions, taking into consideration all conditions affecting the operation.

LAYING UP THE BOILER

A boiler needs the protection of water treatment when it is in service, but there are occasions when a boiler is out of service—for example, during the summer—or other periods of low steam demand. When out of service, a boiler needs just as much protection from the damage caused by corrosion as when in service.

There are two methods of "laying up" a boiler, that is preserving a boiler when it is idle. One is called a wet layup and the other is a dry layup. Both methods are used, depending on the circumstances and the duration of the layup.

For short periods, up to 3 months, the wet system is adequate; for longer periods a dry layup is recommended.

For both methods the boiler must be thoroughly cleaned on both fire and watersides. Before cleaning, close all outlet valves from the boiler.

Wet Layup

For wet storage, after cleaning, fill the boiler to a half gauge and start the fire, raising the temperature of the water until steam issues from the air vent. Add sufficient tri-sodium phosphate to raise the total alkalinity to 400 ppm. Add sufficient sodium sulfite to raise the sulfite content of the boiler water to 5 ppm. The chemical additives will prevent acidic corrosion and oxygen pitting. When the chemical composition is correct, add water to the boiler until the boiler is full. When water starts to issue from the mainsteam stop-drain valve, close the drain and the air vent. Allow the pressure in the boiler to rise to 5 to 10 psi and then shut off the fire. The boiler will cool down and a vacuum will be created in it. No air should be permitted to leak into the boiler; to confirm this, attach a vacuum gauge to the air vent. Periodic checking of the vacuum reading will confirm that the boiler is airtight. If air leakage does develop and the vacuum is lost, check the boiler water occasionally to confirm that the alkalinity and sulfite levels are still adequate to protect the boiler. The fireside dampers and man doors should be closed, and if possible, a canvas or metal covering should be placed over the stack. If this is done, the boiler controls should be posted, advising of the stack covering. Lighting off a boiler with the stack cover on will result in a boiler room full of smoke and the possibility of a boiler explosion.

Dry Layup

To lay up a boiler in the dry state, which is sometimes a necessity in cold climates during a plant shutdown period, the preparation is the same as for a wet layup, except for filling the boiler with water.

Instead of filling the boiler, after cleaning, force air with a blower through all water spaces, then place trays of burning charcoal in the water side of the boiler and close up the boiler airtight.

The burning charcoal will absorb all oxygen in the boiler, thus preventing oxidation.

SETTING UP A SHIFT SCHEDULE

A boiler plant in large industrial or utility operations is usually a 7-day 24-hour operation, the year round. In order to provide for continuous coverage of the boilers it is necessary to establish a shift schedule for operating personnel. Assuming a typical 8-hour day, the schedule illustrated in Fig. 10-3 allows for a three-shift operation with four

OPERATING ENGINEERS SCHEDULE

Week of _____

Watch	Mon	Tues	Wed	Thurs	Fri	Sat	Sun
12:00 MN - 8:00 AM	A	A	A	A	A	A	D
8:00 AM - 4:00 PM	B	B	C	C	C	C	C
4:00 PM - 12:00 MN	D	D	D	D	B	B	B

Week of _____

Watch	Mon	Tues	Wed	Thurs	Fri	Sat	Sun
12:00 MN - 8:00 AM	D	D	D	D	D	D	B
8:00 AM - 4:00 PM	C	C	A	A	A	A	A
4:00 PM - 12:00 MN	B	B	B	B	C	C	C

Week of _____

Watch	Mon	Tues	Wed	Thurs	Fri	Sat	Sun
12:00 MN - 8:00 AM	B	B	B	B	B	B	C
8:00 AM - 4:00 PM	A	A	D	D	D	D	D
4:00 PM - 12:00 MN	C	C	C	C	A	A	A

Week of _____

Watch	Mon	Tues	Wed	Thurs	Fri	Sat	Sun
12:00 MN - 8:00 AM	C	C	C	C	C	C	A
8:00 AM - 4:00 PM	D	D	B	B	B	B	B
4:00 PM - 12:00 MN	A	A	A	A	D	D	D

Figure 10-3. Typical three-shift operating schedule.

people, giving each operator a 40-hour week with the operator on the 12–8 shift working 8 hours of overtime each week.

This form of scheduling is called "rotating shift" and provides the opportunity for each operator to work each shift. Other forms of scheduling are possible but the one illustrated provides equal hours of work for all personnel.

EXERCISES

1. What are operating parameters?
2. What are the important entries on a log sheet?
3. How would you establish the cost of producing 1000 pounds of steam?
4. In addition to steam produced, what factors influence water consumption?

5. When opening a boiler, which door should be opened first? Why?

6. Why should you restrict moisture on the fireside of a boiler?

7. What should be the level in a gauge glass when filling a boiler to start raising steam? Why?

8. Why should you have the air vent open when draining a boiler? When raising steam?

9. Why is a boiler laid up dry?

MULTIPLE CHOICE QUESTIONS

1. When starting a plant for the first time, the operator should have a good set of:
   a. Tools
   b. Instructions
   c. Operating parameters

2. Evaluation of equipment performance requires:
   a. Constant supervision
   b. Daily log sheet
   c. Manufacturer's representative

3. If the monthly cost of operating a boiler room is $100,000 and if 30,000,000 lb of steam are produced each month, the cost per 1000 lb of steam would be:
   a. $33.33
   b. $3.333
   c. $333.3

4. A boiler should be internally inspected at least:
   a. Annually
   b. Monthly
   c. Biweekly

5. When lighting up a boiler after internal inspection:
   a. Have another operator present
   b. Tighten doors as boiler warms up
   c. Report to insurance company

6. Laying up a boiler means to shut it down:
   a. For repairs
   b. For a long time
   c. Briefly

7. A wet lay up requires treatment of the _____ against corrosion.
   a. fuel lines
   b. fireside
   c. waterside

8. To prevent moisture corroding the fireside during lay up:
   a. Cover stack
   b. Light the fire
   c. Neither a nor b

9. When laying up a boiler dry, it is necessary to remove oxygen from:
   a. The waterside
   b. The fireside
   c. The air flow

10. A rotating shift is:
    a. A roller gear
    b. A mechanical device
    c. A watch-keeping schedule

Boiler Room Safety

At various points throughout this book, as appropriate, precautionary measures against the dangers inherent in boiler room operations have been suggested. In addition, efficiency of operation has been emphasized. Experience has shown that the most efficient operations are also the safest. The knowledge needed to operate a boiler plant efficiently is reflected in the preventive steps taken against accident and injury.

Except in the larger plants, operating engineers are usually on watch alone, with total responsibility for the equipment in their charge. A thoughtless act, opening the wrong valve, starting the wrong pump, or neglecting the equipment can produce injury to the operator and damage to the power plant. Accidents generally result from human error or an equipment or material defect.

As in all forms of human conduct, a set of rules to live by is the best way to provide for a safe boiler room. These rules should anticipate any human error that would cause impairment to equipment or injury to the operator. The following list, while not complete, suggests the direction of boiler room safety regulations.

BOILERS

(1) When lighting off a boiler make sure that sufficient time has been allowed to purge the boiler fireside clear of all possible explosive mixtures. If boiler is equipped with

automatic controls, never manually advance program to shorten purge time.

(2) Always check boiler water level at the start of each shift by blowing down gauge glass. Check low-water cutout control daily.

(3) When blowing a boiler down, never walk away from the blowdown valve before closing it. It is too easy to be distracted and forget that the valve is open.

(4) When testing safety valves, have a wire rod attached to the lever so that you are a safe distance from the valve.

(5) Always be sure that the drain line from the safety valve is kept clear.

(6) Before using a soot blower, make sure you have dry steam at the valve and that the velocity of air through the boiler is high enough to carry soot particles clear of the breeching and up the stack.

(7) When opening a main stop valve, under pressure, "crack" the valve first. Do not force a valve handwheel by using a lever or other device. When a steam valve is cold, do not tighten in the closed position, except by hand. As the valve warms up when steam is raised, the steam will expand, jamming the valve closed.

(8) If for any reason, you suspect that the boiler is low on water, shut off the fire immediately. If a coal fire, smother the fire with sand or wet fine coal. It is better to lose steam for a short period of time than the boiler for a long time.

(9) Never tolerate any kind of oil, gas, or water leak around the boiler. If a leak develops, repair it promptly. You don't know what will happen if the leak is neglected.

(10) When opening a boiler for cleaning or inspection, make sure boiler is vented, by opening the air cock.

(11) Always open the top door on boiler waterside first.

(12) When working inside a boiler always have a man outside the boiler watching you. Before entering a boiler, thoroughly vent with forced air.

(13) When a boiler is being cleaned, lock all valves into the boiler closed or blank off or disconnect the lines. If there is a direct connection to another boiler or steam main, open a drain valve to blow to atmosphere on that connection.

(14) Lock out all electrical switches and post a sign on the front of the boiler advising that boiler is being worked on.

(15) Before closing a boiler, make sure all tools, and other articles are removed from all spaces.

PUMPS AND HEATERS

(1) Never run a pump with suction and discharge valves closed. Overheating of the pump and damage will result.

(2) Make sure all coupling guards are in place before starting a pump.

(3) If it is necessary to reduce the flow or throttle a pump, use the discharge valve and not the suction valve.

(4) Make sure that all positive displacement pumps have a relief valve on discharge piping ahead of a shutoff valve.

(5) When an electric fuel heater has a no-flow condition, shut off the power to the unit. Overheating of the oil could cause an explosion and fire.

FIRE SAFETY

(1) Good housekeeping is the first rule of boiler room fire safety. Do not allow any accumulation of rags, trash, or waste oil.

(2) Clean up any oil or water spills immediately.

(3) Know the location of fire extinguishers and how to operate them.

(4) All fire extinguishers in a boiler room should be the carbon dioxide or dry-chemical type, for use on Class A, B, and C fires.

(5) All boiler rooms should be equipped with a remote electric power shut off switch to shut down fuel oil pumps outside the boiler room.

EXERCISES

1. Why do we need safety rules in a boiler room?

2. What is a boiler purge?

3. What is the first thing an operator should do when starting his/her shift?

4. Why is it important not to walk away from the valve when blowing a boiler down?

5. If a pump has to be throttled, which valve should be used?

6. Why is it important to check inside a boiler before closing?

7. What type of fire extinguisher is suitable for a boiler room?

8. What is the purpose of a relief valve on the discharge of a pump? What type of pump needs a relief valve?

MULTIPLE CHOICE QUESTIONS

1. Accidents in a boiler room are usually caused by:
   a. High pressure
   b. Human error
   c. The wrong tools

2. The most important factor in preventing boiler fireside explosions is:
   a. Purge time
   b. Good operators
   c. Boiler room exhaust fans

3. Which type of fuel should have a longer purge time?
   a. Gas
   b. Oil
   c. Neither

4. If you cannot find water in a boiler, you should:
   a. Add more water
   b. Call a supervisor
   c. Shut down the boiler

5. The first act when taking over a boiler watch is to:
   a. Check the steam
   b. Clean strainers
   c. Check the water in the boiler

6. Water can be added to a boiler when the highest heating surface is covered with water. You can check this with the:
   a. Gauge glass
   b. Try cock
   c. Blowdown valve

7. Before draining a boiler:
   a. Make sure it is cool
   b. Make sure there is no pressure
   c. Make sure air cock is open

8. Fire extinguishers in a boiler room should be:
   a. Foam
   b. Soda acid
   c. Classes A, B, and C

9. A positive displacement pump needs:
   a. A strong motor
   b. A relief valve
   c. Purging

10. A remote electrical switch will allow you to:
    a. Operate the boiler room from a distance
    b. Shut off power to the boiler room
    c. Electrically switch pumps

Glossary

Absolute Pressure: Gauge pressure plus barometric pressure.

Actuator: A controlled motor, in which electric energy is converted into action.

Air Flow Switch: A control used to indicate air movement through a combustion chamber.

Air-Fuel Ratio: The ratio of air supply flow rate to the fuel supply flow rate.

Air Register: A burner mounting which may admit secondary air to the combustion space.

Alarm: An audible or visible signal indicating an abnormal condition.

Ambient Temperature: The temperature of the air and/or other gases immediately surrounding a device.

Annunciator: A device which indicates an off-standard or abnormal condition by visual signals.

Ash: Noncombustible matter that remains after a fuel is burned.

Atmosphere: A unit of pressure equal to 14.696 pounds per square inch.

Atmospheric Burner: A burner in which all air for combustion is supplied by natural draft.

Atmospheric Pressure: The pressure exerted on the earth's surface by the weight of the air.

Atomization: The breaking of a liquid into a multitude of tiny droplets.

Atomizer: A device which, with the assistance of an atomizing medium such as steam or air, breaks liquid fuel into tiny droplets.

Atomizing Air: The air supplied through an atomizing air burner (usually about 10 percent) which is used to break the oil stream into tiny droplets.

Automatic Control: A system that reacts to a change or imbalance in one of its variables by adjusting the other variables to restore the system to the desired balance.

Automatic Ignition: A system in which a burner is ignited directly by an automatically provided spark or by a gas or oil pilot.

Available Heat: The quantity of heat released in a combustion chamber that is available for useful purposes.

Backwashing: The term used for the cleaning of a filter bed. The passing of water through a filter bed in a reverse flow to remove entrapped particulates.

Baffle: A metal or refractory surface used to divert the flow of flame or flue gases.

Barometric Damper: A balanced air valve used to admit air to the flue pipe to maintain a constant draft through a furnace.

Bimetallic Element: A device activated by the difference in the coefficient of expansion of dissimilar metals when bonded together.

Boiler: An appliance used to supply hot water or steam for heating, processing, or power generation.

Boiler Horsepower (bhp): The equivalent of the heat required to change 34.5 pounds of water per hour at 212 °F to steam at 212 °F. It is equal to a boiler heat output of 33,475 Btu per hour.

Bourdon Tube: A tube that responds to pressure changes. The tube is shaped into an arc or spiral with one end attached to an indicating device; an increase of pressure within the tube makes it less elliptical, thus actuating the device.

Breeching: A duct to conduct flue gases from the furnace or boiler to the chimney.

Brine Softening: The removal of calcium and magnesium ions from water in exchange for sodium ions.

British Thermal Unit (Btu): The quantity of heat required to raise one pound of water one degree Fahrenheit.

Burner Input Control Valve: An automatic control valve for regulating the input of fuel to a burner.

Butterfly Valve: A throttle valve consisting of a disk which rotates about its diameter.

Centrifugal Atomizing Oil Burner: A burner in which oil is thrown by centrifugal force from a rotating cup into an air stream.

Circulating Loop: The main loop in which oil is circulated from storage tanks to branch circuits and then back to the storage tank.

Combustion: The rapid oxidation of fuel gases accompanied by the production of heat and light.

Combustion Air: The air required for complete and satisfactory combustion of the fuel.

Combustion Chamber: The portion of the heating or process equipment where fuel is burned.

Combustion Control Valve: A control valve for regulating burner input.

Compound Pressure Gauge: Gauge constructed to read both pressure and vacuum.

Compressed Air: Air at a pressure above atmospheric pressure.

Condensate: The liquid which separates from a gas due to a reduction in temperature.

Conduction: The transfer of heat through a material by passing it from molecule to molecule.

Control Valve: The valve which regulates the fuel flow to a burner with variable firing rate.

Convection: Transfer of heat by moving masses of matter. Convection currents are set up in a fluid because of differences in density at different temperatures.

Cross-Connected: Two or more pipes or systems of flow connected to each other.

Crude Oil: Unrefined oil in its natural state as it comes from the ground.

Damper: A valve or plate for controlling draft or flow of the flue gases. Dampers may be automatically or manually operated.

Deaeration: The removal of noncondensible gases from water.

Demineralization: The removal of soluble salts from water using ion-exchange resins. The process of de-ionization.

Draft: The movement of air into and through a combustion chamber, breeching, stack, and chimney. Draft may be natural resulting from the difference in density of the heated air rising through the stack and the cooler displacing air. Artificial draft may be provided by mechanical means.

Draft Loss: The reduction of draft intensity caused by resistance to flow of flue gases through the boiler, breeching, and chimney.

Draft Regulator: A device which acts to maintain a desired draft by automatically controlling the chimney draft.

Dryback Boiler: A boiler where the back of the combustion chamber is lined with refractory brick.

Dual-Fuel Burner: A burner using one prime fuel, but capable of using a standby fuel under peak load conditions.

Electric Ignition: The pilot or oil flame is ignited by electric spark.

Excess Air: Air is supplied in addition to the quantity required for complete combustion.

Filtration: The removal of suspended solids by passage through a porous medium.

Fire-Box Boiler: A boiler with a combustion chamber of cubical design generally made of builtup fire brick (sometimes steel).

Fire Point: The minimum temperature at which a flame is sustained.

Fire-Tube Boiler: A boiler in which the hot flue gases pass through tubes surrounded by water.

Firing Rate: The rate at which air, fuel, or an air–fuel mixture is supplied to a burner or furnace, expressed in volume, weight, or heat units.

Firing Rate Control: A control that automatically regulates the firing rate.

Flashback: The phenomenon which occurs when a flame moves back through a burner nozzle (and possibly back to the air–fuel mixing point).

Flash Point: The minimum temperature at which sufficient vapor collects above a liquid oil for it to catch fire if exposed to open flame.

Flow Rate: The fuel input to a burner measured in suitable units, e.g., cubic centimeters per minute, gallons per hour, pounds per hour. Also referred to as the *firing rate.*

Flue: A passageway for conveying the combustion products to the outer air.

Flue Pipe: The conduit connecting an appliance with the vertical flue of a chimney.

Flue Gases: Gaseous products of combustion.

Flue Gas Loss: The heat carried away by the flue gas and by the water vapor in the flue gas. Also referred to as *stack loss*.

Forced Draft: Mechanically produced air flow into and through the combustion chamber created by a fan or blower located at the inlet air passage to the furnace.

Forced Draft Fan: A fan or blower that forces air into the combustion chamber.

Fuel: Any substance used for combustion.

Furnace Pressure: The gauge pressure in a furnace combustion chamber. The furnace pressure is said to be positive if greater than atmospheric pressure, negative if less than atmospheric pressure.

Gas–Oil Burner: A burner capable of burning either gas or oil.

Gas Pressure Regualtor: A device for controlling and maintaining a uniform outlet gas pressure.

Gauge Pressure: The pressure reading above atmospheric pressure.

Gravity Feed: The oil feed to a burner supplied by gravity from an overhead tank.

Header: A manifold or supply pipe to which several branch pipes are connected.

Heat Content: The total of latent and sensible heat stored in a substance.

Heat Exchanger: Any device for transferring heat from one medium to another.

Heating Value: The heat released by combustion of a unit quantity of a fuel, measured in calories or Btu's.

High Fire: The input rate of a burner or combustion chamber at or near maximum.

High-Low Fire: Provision in a burner for two firing rates, high and low, varied according to load demand.

High-Pressure Hot-Water Boiler: A boiler furnishing hot water at temperatures in excess of 250°F or at pressures in excess of 160 psi.

High-Pressure Steam Boiler: A boiler furnishing steam at pressures in excess of 15 psi.

High-Pressure Switch: A device to monitor liquid, steam, or gas pressure and arranged to shut down the burner at a preset high pressure.

Horsepower: A unit of power equal to 550 foot-pounds per second or 33,000 foot-pounds per minute.

Ignition Temperature: The minimum temperature at which combustion can be started.

Impeller: A series of rotating blades or vanes. In an open impeller, the impeller blades rotate between the stationary walls of the blower housing. A closed impeller has cover plate disks attached to the sides of the blades.

Incomplete Combustion: Combustion in which fuel is only partially burned, and is capable of being further burned under proper conditions.

Induced Air: Air which flows into a furnace because the furnace pressure is less than atmospheric pressure.

Induced Draft: Mechanically produced air movement into and through the combustion chamber and breeching, creating a pressure sufficient to exhaust the combustion products.

Induced-Draft Fan: A fan or blower that produces a negative pressure in the combustion chamber.

Interlock: An automatic control which (1) proves that conditions for combustion are established, or (2) proves that the burner is ready to be started.

Interrupted Pilot: A pilot automatically lighted for each burner start, and cut off automatically at the end of the trial for ignition of the main burner.

Lead Sulfide Photocell Flame Detector: A photocell using lead sulfide as the sensing material.

Light Fuel Oil: Grades 1 and 2, which are distillate fuel oils.

Light-Off: The procedure of igniting a burner or system of burners.

Limit Control: A control which continuously monitors conditions.

Liquefied Petroleum Gas (LPG):   Fuel gases, including commercial propane or butane.

Low Fire: A term meaning that the input rate to a burner is at the minimum.

Low-Fire Start: The firing of a burner with the fuel controls in a low-fire position to provide safe operating condition during light-off.

Low-Oil Temperature Switch: A temperature actuated device arranged to effect the safe shutdown of an oil burner or prevent it from starting when the fuel oil temperature falls.

Low-Pressure Hot-Water Boiler: A boiler furnishing hot water at pressures not exceeding 160 psi and at temperatures not above 250 °F.

Low-Pressure Steam Boiler: A boiler furnishing steam at pressures not more than 15 psi.

Low-Pressure Switch: A device to monitor liquid, steam, or gas pressure.

Low-Water Cutoff: A device to effect a safety shutdown and/or lock out the burner when water level falls to a low level.

Lubricated-Plug-Type Valve: A plug-and-barrel-type valve designed for maintaining a lubricant between the bearing surfaces.

Main Fuel Valve: The valve which controls the fuel input to the burner.

Main Stop Valve: The valve connected directly to the boiler which allows steam to leave the boiler.

Makeup Water: Water which is added to a boiler, tank, or some other container to replace water which has been lost, thus maintaining the proper water level.

Manometer: A U-tube containing water or mercury. Used to indicate low pressures above or below atmospheric pressure.

Manually Lighted Burner: A burner in which fuel to the main burner is turned on only by hand.

Manual Reset: The manual operation required after safety shutdown before the unit can be restarted.

Manual Reset Valve: A valve opened manually and held in the open position electrically or pneumatically or closed automatically by the control system.

Manual System: A system which is purged, started, ignited, modulated, and stopped manually.

Mechanical Atomizer: A device which breaks fuel up into tiny droplets without using an atomizing medium.

Mechanical Draft: The pressure difference, created by machinery such as a fan or blower, sufficient to supply all, or part of, the required combustion air.

Mechanical Draft Burner: A burner designed for use with mechanical draft.

Mechanical Pressure Atomizing Burner: A burner in which oil under pressure is permitted to expand through a small orifice.

Metering Valve: A valve for regulating the fuel input rate to a burner.

Minimum Firing Rate: The lowest input rate for a burner.

Modulating Control: A system of control that provides an infinite number of control positions to fire the burner.

Motorized Value: An automatic valve which is completely opened by the rotation of an electric motor.

Multipass Boiler: A boiler in which the flue gases are passed back and forth through the boiler shell two or more times by means of a tube arrangement or baffles.

Natural Draft: The pressure difference resulting from the tendency of hot gases to rise up a vertical flue, or chimney, thus creating a partial vacuum in the furnace.

Natural Draft Burner: A burner dependent upon the effect of natural draft for the combustion air supply.

Natural Gas: A gaseous fuel occurring in nature, consisting mostly of a mixture of organic compounds (normally methane, butane, propane, and ethane). The Btu value of natural gases varies between 700 and 1500 Btu per cubic foot, the majority averaging 1000 Btu.

Nozzle-Mixing Burner: A burner in which the fuel and air are not mixed until just as they leave the burner port.

On-Off Control: A two-position control which turns the input on or off.

Open-Type Burner: A burner surrounded by an air register through which secondary air can enter the furnace.

Operating Range: The region between the maximum fuel input and minimum fuel input in which the burner flame can be maintained.

Orifice: An opening or construction in a passage to regulate the flow of fuel to a burner, or steam flow through a pipe.

Orifice Plate: The orifice inserted between flanges to measure steam or water flow via the differential pressure across the plate.

Overfire Draft: The draft measured over the flame within the combustion chamber.

Overrate Firing: Firing a boiler at an input rate in excess of its rated capacity.

Oxidizing Flame: A lean flame or fire resulting from combustion of a mixture containing too much air.

Package Generator: A steam or hot-water boiler in which the pressure vessel, furnace, or burner is designed, assembled, wired, and shipped as an integral unit.

Percent Excess Air: The percentage of air supplied in excess of that required for complete combustion.

Perfect Combustion: The combining of the chemically correct proportions of fuel and air in combustion so that the fuel and oxygen are both totally consumed.

Pilot: A small burner which is used to light off the main burner.

Pilot Supervision: Detecting presence or absence of a pilot flame.

Plug Cock: A hand-operated valve to close off the gas supply to a manifold.

Postpurge: A period of time after fuel valves close during which the burner motor or a fan continues to run.

Pour Point: The minimum temperature at which fuel oil can be pumped or flows readily.

Power Burners: A burner in which all air for combustion is supplied by a motor-driven fan.

Preheated Air: Air heated prior to its use for combustion.

Premix Burner: A burner in which all or most of the air required for combustion is mixed with the fuel prior to discharge at the burner ports.

Prepurge: A period of time on each startup during which a burner motor or fan runs to change the air of the combustion chamber and breeching prior to an attempt to ignite.

Pressure Atomizing Gun-Type Burner: Burner where light oil (up to 300 lb/in²) is discharged through a small nozzle or orifice.

Pressure Regulator: A device used to maintain a constant pressure in a supply line regardless of the flow.

Pressure Switch: A pressure-responsive switch.

Pressure Vessel: The steel drum or collector where generated steam or hot water is accumulated.

Primary Air: Air which is mixed with the fuel prior to ignition or burning of the mixture.

Primary Control: A control which provides a means for starting the burner in the proper sequence, proving that the burner flame is established and supervising the flame during burner operation.

Propane: An easily liquefiable hydrocarbon gas. Propane is one of the components of raw natural gas.

Purge: Eliminating a substance from a pipe or furnace by flushing it out with another substance.

Recycle: The process of sequencing a normal start following shutdown.

Refractories: Heat-resistant materials used to line furnaces, kilns, ovens, and combustion chambers.

Refractory Block: A piece of refractory material molded with a conical or cylindrical hole through its center and mounted so that the flame fires through this hole.

Register: The porting, grill work, or damper arrangement through which air is introduced to the combustion chamber around the burner ports or nozzles.

Relief Valve: A valve which opens at a designated pressure and bleeds a system in order to prevent a buildup of excessive pressure.

Safety Control: Automatic controls and interlocks (including relays, switches, and other auxiliary equipment used to form a safety control system) which are intended to prevent unsafe operation of the controlled equipment.

Safety Shutoff Valve: A valve that is automatically closed by the safety control system or by an emergency device to completely shut off the fuel supply to the burner.

Safety Switch: A time delay device which locks out on safety shutdown and must be manually reset before burner can restart.

Scanner: A term applied to all flame detectors.

Secondary Air: Air which is supplied to the flame at the point of combustion.

Set Point: A predetermined value to which a control or interlock is adjusted.

Shutoff Valve: A manually-operated valve in a fuel steam or water line.

Solenoid Valve: An automatic valve which is opened or closed by the action of an electrically excited coiled wire magnet upon a bar of steel attached to the valve disk.

Steam Atomizing Burner: A burner which uses high-pressure steam to assist in atomization of the fuel oil.

Theoretical Air: The chemically correct amount of air required for complete combustion of a given quantity of a specific fuel.

Thermocouple: A pair of conductors joined at their ends. The heating of one end generates an electromotive force.

Thermostat: An automatic control actuated by temperature change to maintain temperatures within predetermined limits.

Throttling Valve: A valve used to control the flow rate of a fluid.

Torch Ignition: Provision for manually igniting a pilot or main fuel valve.

Turndown: The ratio between maximum usable flow and the minimum controllable flow.

Vent Valve: A normally open valve piped between the control valve and final shutoff valve and vented to the atmosphere.

Viscosity: A measure of the flowability of a liquid at a definite temperature.

Water Leg Boiler: A boiler generally of fire-box design where the water vessel extends to the base on both sides of the combustion chamber.

Water Tube Boiler: A boiler where the water being heated passes through tubes surrounded by the flue gases.

Wetback Boiler: A boiler where the water vessel extends down the back of the combustion chamber.

Zero Gas: Gas at atmospheric pressure (zero gauge pressure).

Zero Governor: A regulating device which is normally adjusted to deliver gas at atmospheric pressure within its flow rating.

Index

A

Acceleration, 147
Accumulation test, 109
Adamson ring, 71, 72
Adiabatic calorimeter, 47, 48
Air, excess, 57, 100
Alkalinity, 194, 195
  methyl orange, 195
American Petroleum Institute,
  45
American Society of Mechani-
  cal Engineers, 78, 80, 82
American Society for Testing
  Materials, 80, 82
Analyses:
  flue gas, 57
  Orsat, 57
  proximate, 38
  ultimate, 38, 39, 40
Analyzer, Fyrite, 57, 58
Anion beds, 212

Anthracite, 38
Armstrong, Adam, 180
Armstrong Machine Works,
  180
Arner, Otto, 180
A.S.M.E. Code, 109
Ash, 44
Atmospheric pressure, 3, 6, 9
Atomization:
  centrifugal, 129
  mechanical, 129
  steam-assisted, 129
Avogadro, 54
Avogadro's law, 54

B

Babcock & Wilcox, 78
Babcock & Wilcox "D" type
  boiler, 93

Babcock & Wilcox F.M. package boiler, 91
Babcock water tube, 77
Backwashing, 204
Barometer, 3
Beam engine, 6, 14
Bimetallic element, 220
Blowdown, 109, 110
  boiler, 126
  care of, 93
  heat recovery, 128
  tank, 127
Boiler:
  blowdown, 126
  care of, 93
  cleaning, 241
  closing, 242
  Cornish, 68, 69, 73
  cutting in, 243
  dry back, 73
  fire box, 86
  four-pass, 89
  haystack, 67, 71
  horizontal return tube, 75
  horsepower, 134
  inspection, 239
  Keeler Type CP, 92
  Lancashire, 68, 73
  lay up, 243
  pressure vessel code, 78
  room log, 238
  safety, 249
  Scotch marine, 75, 77
  start up, 242
  Stirling, 77, 93, 96
  Three-pass Atlas, 76
  Tri fuel, 133
  stress in shells, 79, 80
  vertical, 74
  wagon, 67, 69
  water control, 228
  wet back, 86
  Yarrow, 77
Bottom sediment and water, 44
Bottom sludge and water, 166
Boulton, Matthew, 9, 10, 11
Bourdon tube, 223
Bowling hoop furnace, 71, 72
Boyle, Robert, 4
Boyle's law, 26, 28
Brake horsepower, 147
Brine softening, 193
British thermal unit, 20, 21, 22
B.T.U. meters, 226
Burners, 129
  dual fuel, 133
  nozzle mix, 130
  output, 51
Butane, 53
Butterfly gas valve, 232

C

Calcium carbonate, 192
Calley, John, 4, 6
Calorific value, 45
Carbon 38, 43
Carbon dioxide, 54–55
  corrosion, 199
  granulated, 203
Carbon monoxide, 57, 100
  analyzer, 100
  control, 233
Carboniferous Period, 37
Carryover, 95
Cation beds, 212
Cavitation, 146

Celsius, Anders, 18
Celsius scale, 18
Centigrade scale, 18
Charles, Jacques, 27, 28
Check valve, 115
Chloride, 195
  calcium, 196
  magnesium, 196
  sodium, 196
Chloride test, 196
"Claremont," 75
Coal, 37, 38
  ash content, 39
  bituminous, 38
  burning, 132
  gas, 53
  sizes, 38
Coefficient, valve sizing, 170
Coefficient of expansion, 25
Combustion:
  chamber, 88
  control, 230
  efficiency, 60, 61, 62, 63
  process, 43
  products of, 54
Condensate return header, 181
Condensation, 29
Condenser, tubular, 12
Conduction, 23
Constant heat expansion, 169
Controls, 219
Convection, 23
Corrugated furnace, 73
Crown sheets, 89

D

Da Vinci, Leonardo, 144
Dark Ages, 2

Deareation, 207
  mechanical, 197
De Caus, Salomon, 2
Della Porta, Giovanni Battista, 2, 3
Degrees:
  absolute, 18, 19
  centigrade, 18
  Fahrenheit, 18
  Reaumer, 18
Demineralization, 193, 209, 211
  mixed bed, 212
  regeneration, 212
Disc trap, 179
Dissolved oxygen, 197
Dissolved solids, 191
Drake, Colonel, 41
Dry lay up, 244
DuLong, Pierre Louis, 40
DuLong's formula, 40

E

Ebullition, 28, 29
Ecole Polytechnique, 40
Efficiency:
  boiler, 56, 98, 99
  combustion, 60, 61, 62, 63
  engine, 10
  improvement, 100
  of operation, 239
  operational, 99
Effluent blending, 213
Engine:
  atmospheric, 7
  beam, 6, 14
  condensing, 9, 67
  Newcomen, 8, 9

Engine (*cont'd.*)
  pumping, 10
  steam, 9
Enthalpy, 32
Environmental Protection
    Agency, 43
  regulations, 39
Ethane, 52
Evaporation, 28, 29
Expansion:
  areal, 25
  coefficient of, 26
  linear, 25
  volumetric, 25
External treatment, 202

F

Factor of safety, 81
Fahrenheit, Gabriel Daniel,
    17
Feedwater:
  degassing, 207
  heater, 171
  injector, 136
  piping, 135
  quantity, 135
  systems, 133, 134
Filters:
  gravity, 203
  horizontal, 205
  operation of, 206
  pressure, 205
  sand, 203
  types of, 203
  vertical, 205
Filtration, 202
Fire point, 43, 51
Fire safety, 251
Fireside, 97

Flash point, 43, 51, 52, 164
Float thermostatic trap, 179
Flocculation, 203
Flocculents, 94
Flow measurement, 225
Flow meters, 228
Flow restriction, 170
Foot pound, 21
Forced draft fan, 230
Friction head, 145
Fuel oil:
  flash point, 43, 164
  heaters, 168
  heating value, 26
  pour point, 43
  pump, 168
  specific gravity, 43, 45
  storage, 52
  strainers, 166
  systems, 167
  viscosity, 43, 164, 165
  volume corrections, 48
Fuels, 37
  calorimeter, 40, 41
  chemical composition
  economics, 41
  gaseous, 37
  heating value, 48
  liquid, 37
  oils, 41
  solids, 37
Fulton, Robert, 75

G

Galileo, 3, 144
Galloway tube, 68
Gas, 52
  combustion, 53

heating value, 53
modulating cam, 232
natural, 52
pilot adjustment valve, 232
pilot valve, 231
pressure limit switches, 232
pressure regulator
raw, 52
Gases, 25
noncondensable, 208
Gaskets, 97, 98
Gate valve, 115
Gauge:
  compound, 166
  glass, 111, 112, 114
Gay Lussac, Joseph, 27
Gearing, sun-and-planet, 11,
  14
Globe valve, 115
Grand Duke of Tuscany, 3
Gussets, 67

H

Hardness, of water, 195
Hardness test, 195
Heat, 17, 20
  conversion, 23
  effects of, 24
  latent, 22, 179
  measurement, 17
  mechanical equivalent of,
   21
  of condensation, 22
  of fusion, 22
  of melting, 22
  sensible, 179
  solar, 20
  total, 22, 32

transfer, 23, 90
of vaporization, 22
Heater, deareation feed, 174
  feedwater, 171
Heating value, 55
  coal, 40
  fuel oils, 46
Hero of Alexandria, 1
High-pressure limit, 230
High suction, 164
Hooke, Robert, 4
Horsepower, 10, 21
Hot phosphate softening, 193
Hydrogen, 38, 43
Hydrometer, 45

I

Ignition transformer, 230
Industrial Revolution, 11
Instruments, 219
Inverted bucket trap, 179, 180
Ion exchange, 209

J

Jet contraction, 170

K

Keeler CPM generator, 90, 92

L

Lap joints, riveted, 82, 83
Latent heat:
  of condensation, 22

Latent heat (*cont'd.*)
  of fusion, 22
  of melting, 22
  of vaporization, 22, 31
Level control, 228
Lignins, 194
Lime soda softening, 193
Liquid flow, 227
Liquid measurement, 227
London, 4
Low-pressure limit, 230
Low suction, 164
Low water cut off, 230

M

Magnesium hydroxide, 193
Manometer, 224
Mechanical burner, 129
Methane, 52
Modulating pressure control,
  230
Modulating motor, 230

N

Net positive suction head, 144
Newcomen, Thomas, 4, 6, 7,
  67
Nitrogen, 43
Nonreturn valve, 118, 119

O

O rings, 159
Ohms per centimeter, 214

Oil:
  crude, 42
  fuel, 41
  storage, 163, 164
  systems, 163
One-element control, 229
Operating pressure control,
  230
Orsat analyses, 57, 232
Oxidation, 53
Oxygen:
  corrosion, 93
  free, 100
  pitting, 93

P

pH, 104, 194, 198
Penskey-Martin Closed Cup,
  51
Petroleum, origin of, 42
Phosphate:
  monosodium, 194
  tricalcium, 194
  trisodium, 194
Pitot tubes, 226
Plug cock, 121
Power plant management, 237
Pour point, 43, 51
"Potter Cords," 6
Pressure:
  atmospheric, 3, 9
  differential, 171
  drop, 171
  filter, 205
  filter piping, 206
  of gas, 27
  gauge, 225

maximum allowable, 109
measurement, 223
saturation, 170
vapor, 29
Programing control, 230
Propane, 53
Pump, 139
   centrifugal, 139
   duplex, 139
   efficiency, 146
   friction, 148
   gland cooling, 157
   multistage, 140, 143
   packing, 149
   positive displacement, 139, 168
   problems, 160
   rotary, 140
   safety, 251
   sealing, 153
   steam, 4, 139
   volute, 149

R

Radiation, 24
Raising steam, 242
Reaumer scale, 18
Reducing valve, 169
Reducing valve sizing, 170
Resistance pyrometer, 222
River Thames, 4
Riveted joints, 82, 83, 84, 85
Rotary burner, 129

S

Safety valve:
   blowdown, 109, 110
   dead-weight, 105, 106
   maintenance, 110
   spring-loaded, 105, 107
   sizing, 110
Salts:
   calcium, 194
   magnesium, 194
Sampling bomb, 166
Saturation pressure, 170
Saybolt Furol scale, 47, 51
Saybolt Universal scale, 47, 51
Savery, Thomas, 4, 67
Scale, loss of efficiency, 192
Scale formation, 94, 192
Scotch marine boiler, 88, 89
Sedimentation, 203
Shearing strength, 86
Shift schedules, 244
Silica scale, 193
Sizing, steam trap, 186
Sodium:
   sulfate, 197
   sulfite, 197
   zeolite, 209
Softening, 209
Soot blowers, 125
Specific conductance, 214
Specific gravity, 43, 45
Specific heat, 22, 52
Stack gas measurement, 232
Static head, 144
Static lift, 144
Stay bolts, 88, 96
Stays:
   girder, 88, 89
   through, 88
   tubes, 88
Steam:
   dry saturated, 30

Steam (*cont'd.*)
  flow, 225
  powered fountain, 3
  production graphs, 240
  saturated, 29
  superheated, 29, 30
  supersaturated, 29
  tables, 30, 31
  traps, 179
  wet saturated, 30
Stephenson, George, 73, 75
Stirling boiler, 79
Stoker, ram feed, 132
Stokers, 132
Stress:
  circumferential, 80
  longitudinal, 80
  safe, 81
Suction pipe, 4
Sulfur, 43
Sun-and-planet gearing, 11, 14
Surface pyrometer, 188
Suspended solids, 202

T

Tannin, 194
Tell-tale holes, 89
Temperature, 17
  absolute, 18
  conversion, 18
  furnace, 93
  ignition, 53
  kindling, 53
  measurement, 17
  regulating valve, 172
  stack, 56
Tensile strength, 79

Test:
  alkalinity, 190, 195
  chloride, 200
  colorimetic method, 200
  for pH, 202
  hardness, 200
  high-phosphate, 201
  kit, 195
  sulfide, 201
  sulfite, 197
Thermocouple, 221
Thermodynamics:
  First Law, 21
  Second Law, 21
Thermostat, 169
Thermostatic air vents, 182
Thermostatic bellows trap,
    184
Thermostatic trap, 179
Three-element control, 229
Total dissolved solids, 198
Total dynamic head, 145, 146
Total heating surface, 90
Torricelli, Evangelista, 3
Trap:
  bypass, 186
  disc, 179
  drip, 185
  float thermostatic, 179
  inverted bucket, 179
  orifice, 179
  sizing, 186
  test valve, 187
  thermostatic, 179
  trouble shooter, 188
Traveling grate, 230
Try cock, 114
Tube cleaning, 98
Tubes, 97

Tube sheets, 97
Turbine meters, 226
Tuscany, Grand Duke of, 3
Two-element control, 229

U

U. S. Bureau of Mines, 45
U. S. Bureau of Standards, 45

V

Vacuum, 3, 6
Valve:
  ball, 121, 122, 123
  butterfly, 232
  check, 115
  check horizontal, 124
  check vertical lift, 123
  deadweight, 105, 106
  drip leg, 184
  gas pilot, 232
  gate, 115
  gate nonrising stem, 120
  globe, 115
  main gas, 232
  nonreturn, 118, 119
  O.S. & Y., 117, 118
  reducing, 169, 172, 173
  sizing coefficient, 170
  spring-loaded, 105, 107
  velocity, 171
Vaporization, 28
Vapor lock, 52
Vapor pressure, 29, 145, 146
Velocity, 147
  head, 145
Vena contracta, 170
Vent condenser, 209
Vent line, 166
Viscosity, 43, 47
Viscosity scales:
  Engler, 51
  Redwood, 51
Vivani, 3
Von Guericke, Otto, 3
Vortex shedding, 226

W

Wagon boiler, 67
Water
  column, 111, 116
  enthalpy of, 129
  hardness of, 195
  in fuel oil, 44
  treatment, 191
Watt, James, 8, 9, 10, 11, 21, 67
Western Roman Empire, 2
Wet lay up, 244
Wet steam, 95
Whirling aeolipile, 1, 2